Television Technology:

A Look Toward The 21st Century

Television Technology: A Look Toward The 21st Century

Includes selected papers on television technology presented during the 21st Annual SMPTE Television Conference in San Francisco, Calif., February 6-7, 1987.

Foreword
Papers Program Chairman
Joseph Roizen
Telegen

Preface
Editorial Vice-President
Howard T. La Zare
Deluxe Laboratories, Inc.

Editor
Jeffrey Friedman

Editorial/Program Coordinator
Dollie L. Hamlin

Cover Design
Mathew V. Kuriakose

1

Published by the Society of Motion Picture and Television Engineers
595 W. Hartsdale Ave., White Plains, NY 10607

The Society of Motion Picture and Television Engineers,
White Plains, NY 10607

Printed in the United States of America.

Library of Congress Catalog Card Number: 87-60606

ISBN 0-940690-13-6

TELEVISION TECHNOLOGY:
A LOOK TOWARD THE 21st CENTURY

CONTENTS

Page

FOREWORD

The 21st Annual Television Conference, held on February 6 and 7, 1987, in San Francisco, lived up to the tradition set by its predecessors by providing SMPTE members and other television industry delegates with a most comprehensive look at the future of advanced television technology.

When the papers committee for this conference first met in September 1986, it was quickly decided that since this was indeed the 21st Conference and since we were approaching the 21st century, those facts should be utilized as a conference theme. John Streets' suggestions for the title "The 21st Conference Looks to the 21st Century" was readily agreed to, and the sessions were then defined. In the past the San Francisco conference has frequently incorporated new technology displays as part of the overall activities; however, this time there were no really new or significant technological developments that needed to be shown. The papers committee therefore concentrated on having the conference speakers cover three important topics: first, a global overview of the future of television research; second, the impact of that research on major broadcasters and how it will affect their operations between now and the year 2000; and third, an updating on two technologies (computer-assisted graphics and VTR formats) that are both proliferating rapidly and undergoing considerable change.

The late 80s have certainly proved to be watershed years for television technology, and the research developments flowing from the advanced laboratories are constantly being converted from engineering prototypes to practical, operational hardware. Broadcasters and program producers must continue to cope with the steady flow of all this new equipment, even while maintaining the established services that they now supply to both the public and private sectors.

The questions on everyone's mind are pretty much the same, and they relate to the growing dilemma of the multiple pathways to the delivery of television programs in the future: will broadcasters indeed lose their virtual monopoly of the past because they cannot deliver images to the home that compete in picture quality with cable, improved VCRs, fiber optics, or DBS? Will there be two (or more) high-definition television standards based on the current dichotomy between 50-Hz and 60-Hz areas of the world? Will we see such an evolution of the NTSC, PAL, and SECAM color television systems toward such improved pictures in these standards that HDTV will not have the attraction it has today?

The distinguished roster of international speakers at the 21st Conference, who represented the world's leading television research laboratories and the major North American television networks, all provided a comprehensive look at where television technology is headed, and the reprints of their presentations in this SMPTE volume make for very interesting reading. However, the reader will find that, while there are groupings of some unanimity among some of the speakers, the overall situation is one of considerable competition between various points of view as to what is best for the television of the future.

As long as such honest differences of opinion exist, then we can expect that differing television technologies will also continue to coexist and that a universal standard for some single worldwide television broadcasting system or for a single VTR format is only a dream, and one that will not be realized by the end of the 1900s.

Nevertheless, there was much to be grateful for at the 21st Conference. Pockets of important standards do exist, and these serve the industry well. Conversions from one standard to another keep getting better, thus diminishing the degradation of image quality at each incompatible juncture. Last, but certainly not least, the steady overall improvement in the components that go into image origination, recording, and display equipment have all contributed to the high-quality pictures viewers can see, even today, regardless of the delivery system.

If the 21st Conference showed anything, it was the fact that the existing technology already produces excellent television images under most conditions and that the research currently being done on both the technical and psychophysical aspects of television will greatly improve this medium in the near future. That is a message that all of us who work in this field can welcome, and we can look forward with optimism to what the future holds.

As program chairman, I would like to take this opportunity to thank all the members of the San Francisco Section for their invaluable contributions to every phase of the conference, and to the Headquarters staff who provided assistance and support. I would also like to express my appreciation to all of the authors who took the time from their otherwise busy schedules to share their knowledge and experience with all of us.

Joe Roizen, Program Chairman

Joseph Roizen is the president of Telegen, a consulting firm based in Palo Alto, Calif., that provides technical and market research services relating to international television on a worldwide basis. Roizen is a Fellow of the SMPTE, The Royal Television Society, and the BKSTS. In 1984, he was named the International Tape Association "Time Man of the Year." He has been awarded an Emmy Citation for recording the Nixon/Khrushchev Kitchen Debate in 1960 and other industry honors. Roizen has served as an SMPTE Governor, as Conference Program Chairman for the 21st Television Conference, and on the Board of Editors, as well as participating in Standards work. He is the emerging technology editor for Broadcast Television *and a frequent contributing editor to* Television Digest, *the* SMPTE Journal, *the* Journal of the Royal Television Society, *and the* IEEE Spectrum.

PREFACE

The Society of Motion Picture and Television Engineers is pleased to present this collection of papers which were originally given at our 21st Television Conference on February 6 and 7, 1987, at the St. Francis Hotel in San Francisco, Calif. They are replicated in their original form to expedite their publication, and to maintain their timeliness.

The theme of the conference, "The 21st Conference Looks to the 21st Century," was certainly a provocative setting for the discussion of advanced television technology. The papers were grouped under three major topic headings, with the opening dedicated to and entitled "Videographics: The Next Generation." This session featured papers dealing with recent innovations and predictions relating to graphics, animation concepts, and related hardware. The afternoon session was titled "Tape Recording Formats for This Century and the Next." This session, in part, followed SMPTE's highly successful 20th Annual Television Conference, which offered the industry the first definitive description of the SMPTE/EBU Type D-1 Format for digital component tape recording. The papers presented this year continued the discussion of the 4:2:2 recording format, particularly in light of its future potential in broadcasting facilities. Additionally, the session gave consideration to recent developments relative to composite VTR technology and other topics which include the M-II videotape format and the new rotary digital audio tape format (R-DAT). The sessions concluded with an informative panel discussion.

The final day of the conference was a full day of presentations by many of the leaders of the broadcasting industry and television research centers around the world. They assessed today's and tomorrow's technology to give us a glimpse of what tomorrow may hold for us relative to television imaging. The day was highlighted by a panel discussion which included all of the day's speakers. The session was titled "The Frontiers of Global Television Research."

On behalf of the SMPTE, I would like to extend our Society's sincere gratitude to the dedicated group of volunteers and speakers who gave of their time and talent for the membership and community's benefit. The program presentation was under the direction of Program Chairman Joseph Roizen, Telegen, assisted by Program Co-Chairman David Fibush, Ampex Corp., and their supporting committee.

Howard T. La Zare
Editorial Vice-President

Howard T. La Zare, Senior Vice-President, Engineering, Deluxe Laboratories, Inc., is SMPTE Editorial Vice-President. He was most recently SMPTE's Engineering Director, Motion Pictures, and immediately before that, Vice-President for Motion Picture Affairs. Mr. La Zare joined SMPTE in 1971 and was made a Fellow in 1975. He is a member of many committees for the Society. He has previously held the position of Chairman, Manager, and Secretary/Treasurer of the Hollywood Section. Mr. La Zare previously served as Conference Chairman of the 123rd Technical Conference and Program Co-Chairman at the 121st, Session Chairman for the Laboratory Practices Session at the 117th, and Session Chairman and Topic Chairman for the Session on Unconventional Imaging Systems at the 119th. He received, for his engineering achievements, a Class II Academy Award in 1973 and a second Class II Academy Award in 1982 from the Academy of Motion Picture Arts and Sciences.

Opening Address—21st Television Conference

Howard T. La Zare
Deluxe Laboratories, Inc.
Hollywood, California

Good morning Ladies and Gentlemen, Members and Guests. It is indeed a pleasure to welcome you to the Society of Motion Picture and Television Engineers 21st Annual Television Conference.

The theme of the Conference "The 21st Conference Looks to the 21st Century" is certainly provocative, to say the least. During the next two days, you will hear from some of the prime movers of our industries who will give you a glimpse at what these leading engineers and executives consider to be the prevailing television technology of the next 15 years. The papers will be grouped under 3 major topic headings. This morning, "Videographics: The Next Generation"; this afternoon, "Tape Recording Formats for This Century and the Next". And tomorrow, the single topic: "The Frontiers of Global Television Research".

The papers program was orchestrated under the direction of Program Chairman, Joseph Roizen and Co-Program Chairman, David Fibush - and ably assisted by their supporting Committee. Local arrangements were coordinated by General Arrangements Chair Person, Donna Foster-Roizen and her supporting Committee. They deserve our sincere thanks and gratitude for a job well done.

For this opening address, I would like to deviate from the normal opening presentation which usually highlights the events that will take place during the following days of the Conference. This is well stated in your program booklet. Instead, in light of the forecasting and predictions that will be presented during this Conference, I would like to give you a glimpse of what the future holds for us in a much broader view and within the context of the transformation that our nation is going through from an industrial society to an information society and which may be a precursor of what of other recently industrialized nations may look forward to. The years 1956 and 1957 may be considered the turning point and the start of the end of our industrial era.

It had its genesis with the first communication satellite which gave birth to and formed a global community. This technological breakthrough aided by computer technology led to the collapse of the information float--the information float being defined as the time between an occurrence of an incident and the widespread knowledge of it.

Traditionally, we think in economic terms of either goods or services being provided. With most of us no longer manufacturing goods, we assume that the rest of us are providing services, but a close look at the so-called service occupations tells a different story. The overwhelming majority of service workers are actually engaged in the creation, processing and distribution of information. The so-called service sector minus the information or knowledge workers has remained fairly constant equal to about 11 to 12% of the workforce since 1950. The real increase has been in information occupations. In 1950, only about 17% of us were working in information related jobs. Now more than 60% of the workforce work with information, such as programmers, teachers, clerks, secretaries, accountants, stockbrokers, managers, insurance people, bureaucrats, lawyers, bankers, engineers and technicians, with many, many more workers holding information jobs within manufacturing companies. Implying that most Americans spend their time creating, processing or in the distribution of information.

In an industrial society, the strategic resource is capital. A hundred years ago, a lot of people may have known how to build a steel plant, but few could get the capital to build one. Therefore, access to the system was limited, but with information the strategic resource, access to the economic system is relatively easy. A good example of this was the creation of the Intel Corporation - formed in 1968 by Robert Noyce and Gordon Moore, former employees of Fairchild Semiconductor. Intel was started with $2.5 million in venture capital, but it was the brainpower behind the financial resource that led to the technological breakthroughs that brought the firm to annual sales of greater than $850 million by 1980 and is expected to exceed $1.5 billion this year. Noyce is credited with being the co-inventor of the integrated circuit and Intel with developing the microprocessor. With electronic and computer industries being brain rather than capital intensive, it is certainly one of the most important reasons for the current entrepreneurial explosion in the United States. In 1950 we were creating new businesses at the rate of approximately 93,000 per year. Today new companies are being created in excess of 600,000 per year.

The restructuring of our country from an industrial society to an information society was easily as profound a change as was the shift from an agricultural society to an industrial base - with one important difference. It took about 100 years to change our society from an agricultural base to an industrial base. The restructuring from an industrial to an information society took only two decades. With change occurring so rapidly, that there was hardly any time to react. Therefore, we must anticipate the future.

That future is certainly expected to take us to outer space. With the American Government's decision to encourage private launching of satellites and space vehicles, many new U.S. companies will enter this field. Private satellite launching and leasing will boom and not only for communication and broadcasting. More satellites will be put into service for earth mapping, mineral exploration and weather forecasting. In communications, enormous space antennas will provide cellular mobile telephone service over a far wider and vast areas than are now possible.

There will even be small volume, high-priced manufacturing taking place in outer space within privately owned space laboratories and factories. In the gravity-free environment of space, technicians will produce ultra pure drugs, glass, plastics and crystals, free of the microscopic layering and settling effects that are caused by the gravity on earth. Microchip material worth an estimated

12

$50,000 a pound will be produced to meet growing demand for optical switches in the computer industry as well as high-clarity plastic and glass fibers needed in long distance fiber-optic cables.

There is even expected to be the creation of public utilities in space, serving the growing number of commercial space laboratories with electricity generated in space or beamed from Earth via microwave transmissions.

The core technologies of this coming century that will generate the high-technology boom of the future are expected to be in biotechnology, computers, expert systems and artificial intelligence, robotics, lasers and fiber optics, composite materials, batteries and fuel cells. These technologies will represent the building blocks for new products and applications. Working together, they will boost the value of their final products.

The ultimate conveyance of technologies will appear in the form of advanced robotic devices that will "see" with lasers and electro-optical eyes, analyze information with computers and microprocessors, follow instructions from an "expert system" software program verging on artificial intelligence, derive operating power from batteries or fuel cells made of new electric conducting plastics, and have lifting arms built of composites stronger than steel at a fraction of the weight. Combine these advancements with the emerging technology of genetic engineering which will use these advanced robots to decipher and manipulate protein molecules and alter the structure of recombinant DNA. Even today's robots can perform in one day the so-called DNA sequences and synthesize tasks that used to take skilled microbiologists a year to complete.

The manipulation of organic processes and genetic structures will create new products and will be the fastest growing field in the decade ahead. The biotechnologists will create new safer growth hormones to increase the yield of livestocks and crops. New hybrid plants are expected that are more resistant to pests, frost, drought, salty water and chemical build-up and that will even be self fertilizing via a natural nitrogen-fixing process. In human health, bioengineering will produce many breakthroughs from substances that will reduce pain, regulate blood pressure and lubricate arthritic joints to producing new vaccines and drugs that will be effective against most known diseases and many forms of cancer. Applications in energy production will include the creation of microbes that will loosen up oil in seemingly drybed wells and others that will convert tar sand into a usable liquid form which will lower our dependence on imported oil. Add to this the lower cost methods of fermenting alcohol from virtually any vegetable matter including garbage waste. One of the most significant aspects of the bioengineering boom will be the replacement of petrochemicals that are now utilized in many applications such as feedstocks, fertilizers and plastics.

Computers will see enormous advances in speed and compactness. Optical semiconductors in which switching is done by photons rather than electrons will increase today's fastest processing speeds by threefold - approaching the speed of light. Coupling this with further advances in the art of parallel processing and advanced software designs, will make today's computers seem as slow as the mechanical adding machines of yesterday.

"Expert Systems" are just beginning to appear, pulling together vast mounts of knowledge on a given subject, organizing it in such a way that allows computers to analyze problems and reach conclusions via the process of elimination utilizing preprogrammed criteria. These "expert systems" will find their way into most all areas of endeavor, including medical diagnostics, investments and financial planning, teaching, weather prediction, legal strategies based on precedents, geological explorations, to mention a few. "Expert Systems" are the next plateau in artificial intelligence programming. They are the new breed of specialists who combine the talents, in computer programming, with a deep understanding of a particular subject area.

Lasers, those versatile and powerful light sources that are already being used for many tasks such as cutting of steel plates, delicate eye surgery, reading bar codes in supermarkets to deciphering audio and video signals encoded on optical disks are just on the threshold of being exploited. In the near future, lasers will be used to open clogged arteries, thus eliminating many by-pass surgeries. New military defensive systems are being explored. The applications in the home, industry, and defense seem virtually limitless.

In robotics, whether they are welding auto parts, sewing sleeves on suits or delivering office mail, robotic devices will be everywhere in business and homes in the next century. They will plant, irrigate and harvest crops, help put up buildings and raise productivity in countless manufacturing plants. The making, programming and repair of robotic machines may eventually create as many jobs as the devices displace in simple manufacturing.

New materials will increasingly replace traditional materials, such as metals, glass and wood. New composites made of plastics, glass fibers, graphite and ceramics will find countless applications. Ceramic parts are turning up in many industrial uses where resistance to wear corrosion and high temperatures are essential. Ceramic engine blocks will make their debut within the next decade, especially in new gas turbines and adiabatic diesel engines because of their high efficiency and immunity to high temperatures. New polymer complex chemical structures will be adopted for many uses and combined with reinforcing substances. Mixed media polymers will be used in everything from running shoes and airplane propellers to compact discs and low friction ball bearings. Further, ceramic reinforced plastic panels will make their way into home construction and automobiles. A kind of "ultra oriented" polyethylene, six times stronger than steel but much lighter and more resistant to corrosion, will begin to replace steel rods in reinforced concrete construction, especially for bridge decks and roadways. New conducting plastics will appear in solid-state batteries and will be used as molded, built-in wiring inside appliances, without the need for insulation.

Fuel cells will finally come into their own. Their unique capability of acting as chemical reactors generating electricity in disproportionate amounts to cell size, will be fully explored.

It is expected that great strides in power storage technology will take place in the near future leading to compact solid-state batteries capable of making the long-awaited electric automobile practical.

The viable predictions for the next century goes on and on. The thoughts presented today are the amalgamations of thoughts of many learned authors and specialists in related fields. I hope I have opened your eyes and whetted your appetite for the excitement of what waits for us in the near future, and I know that many of you in the audience will make important contributions to bringing these predictions to a reality.

Thank you, and I hope you enjoy the Conference.

Howard T. La Zare, Senior Vice-President, Engineering, Deluxe Laboratories, Inc., is SMPTE Editorial Vice-President. He was most recently SMPTE's Engineering Director, Motion Pictures, and immediately before that, Vice-President for Motion Picture Affairs. Mr. La Zare joined SMPTE in 1971 and was made a Fellow in 1975. He is a member of many committees for the Society. He has previously held the position of Chairman, Manager, and Secretary/Treasurer of the Hollywood Section. Mr. La Zare previously served as Conference Chairman of the 123rd Technical Conference and Program Co-Chairman at the 121st, Session Chairman for the Laboratory Practices Session at the 117th, and Session Chairman and Topic Chairman for the Session on Unconventional Imaging Systems at the 119th. He received, for his engineering achievements, a Class II Academy Award in 1973 and a second Class II Academy Award in 1982 from the Academy of Motion Picture Arts and Sciences.

Today's Videographic Environment—An Overview

Peter C. Lowten
Pixel Consulting & Liaison
Mountain View, California

ABSTRACT:

The utilization of images to enhance information transfer, to entertain and motivate, has created an environment based on two separate approaches to image manipulation, those being via dedicated hardware or general purpose computers.

The interaction of broadcast and business video requirements are modifying manufacturers product directions, as is the groundswell of new creativity being stimulated by the flexible artistic tools that videographic technology has itself created.

These environmental stimuli are encouraging the development of "spinoff" technology which it would otherwise be uneconomic to pursue, and are providing a marketplace for diverse solutions to the manipulation of visal images.

We are each busy with our own niche in life and often do not have time nor opportunity to examine those areas of technology which may not directly affect us today, but surely will before the end of the century. With the enormous build up in information flow now available from many sources comes the need to process, digest and communicate that information Development of methods to present that information in a timely, cost-effective and entertaining manner, along with the creative use of those methods, will impact our lives and careers. Graphical representation is clearly at the forefront of this new direction.

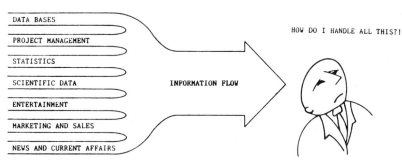

Fig. 1.

It behoves us then to step back and look both at how videographics got this far, and at the current environment from which it must now expand. It is a wide field because videographic techniques are not only being utilized in the entertainment media – broadcast T.V. and films – but increasingly are being put to use in industry and in the multifaceted world of business video, i.e., the use of visual images to market and sell products, to motivate and train employees and end-users, and present accounts, inventories and statistics in such a way as to improve management decisions. With the exception of the last category, broadcast products are already shaped to fit these niches; but business needs demand different cost/performance ratios, quality levels and standards. More and more we will find that the products developed for broadcast use will share in business communications, and more and more that business requirements will enhance new product development. Potentially the business video market is significantly larger than broadcast, and will modify vendors product approaches and R&D efforts. This can be a good trend because spin-off technologies become available to the broadcast marketplace, which would otherwise be too small to justify and motivate their development.

Evolution

The support mechanisms for manipulation and presentation of visual information available today have matured from two very different technologies involved in the acquisition, manipulation and display of graphics materials. These are the technologies underlying Computer-based Systems and Video-based Systems. They have had separate evolutions and now have different, though somewhat overlapping, capabilities derived from fundamentally different starting places.

- Computer-based graphics systems are TEXT and ANALYSIS oriented
- Video-based graphics systems are IMAGE oriented

17

This overview will attempt to relate these approaches, examine the tools now being presented to artists, businessmen and broadcasters and point a direction for the next wave of videographic innovation.

In the context of this discussion, "Videographics" is defined as "graphic image materials developed either from video or computer data sources, by means of video or computer graphic manipulation devices and software".

Visual and Textual Technologies

The ability, driven by varied needs, to acquire and manipulate images has historically evolved from two separate technology areas, one video-based, the other computer-managed data. Although the systems differ, most graphics users will find that the best attributes of both technologies will be needed to support their requirements as we move towards the 21st century. Therefore, it is important to understand the origins and current status of these two types of systems.

Computer graphics came into being when data-based number crunching for computer aided design, development, engineering and management found opportunities for enhancement with graphical tools. These included modeling, drawing, graphing and painting capabilities. The broadcast video industry (with video images as its only concern) was able to increase the sophistication of manipulation equipment based on a single standard of raster scan 525 line video, while the computer industry entered the graphics field as an application of general information processing technology. As a result, while video manipulation devices all handle the same basic types of signals, computer hardware and software vendors have found different ways to use the computer as a platform to produce graphics through the manipulation of data.

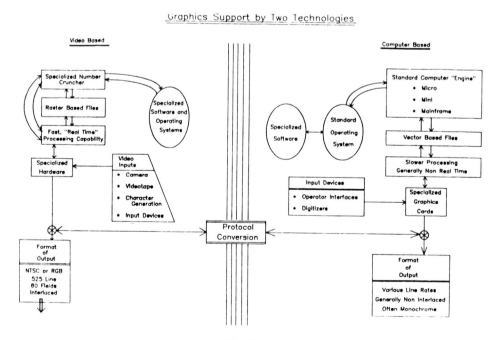

Fig. 2.

This has often hindered the cross-pollination of ideas and product development between the two fields, and broadcast engineers who happen to stray into shows such as the National Computer Graphics Association are often seen shaking their heads and muttering "but what about a broadcast output?" -- the reverse is also true! Real effort on the part of video professionals to achieve liaison between these fields will reduce the wastage of financial and human resources on duplicate and misdirected projects.

It has also led to an even greater stumbling block on the way to full use of graphics capabilities, and that is the integration of higher resolution images into our constrained 525/625 line world. Although compatible today with NTSC video, the continuing High Definition Television (HDTV) research efforts promise significant increases in graphics performance in two main areas. These include the system capabilities themselves, and "spinoff" technology, such as higher resolution cameras and display devices which may be applicable to systems with "downward compatibility", i.e., production of high quality graphics that are easily converted to a format that can be viewed on a standard 525 line monitor. With many competing methods to achieve partial compatibility and expanded capabilities now being offered it seems clear that the acceptance and deployment of high resolution systems will carry us well into the 21st Century.

Creative Tools

We can argue that a videographic presentation consists of:

25% Philosophy, 25% Art, 25% Humanity & 25% Technology.

But we often focus on the technology quarter to the exclusion of the creative talent brought to bear, and are inclined to forget that the hardware tools now presented to artisans using videographics are improving at such a rate that development both of software to utilize them fully, and of creative talent to explore this new canvas, is lagging months and years behind. Far from being a negative attribute this tendency is leading to an explosion of creativity in the broadcast and business video arenas, as imaginative individuals are offered an opportunity to explore a new medium of expression rivaling, and in many ways similar to, the leap from stone carving to drawing on papyrus.

Developing a graphical representation, whether static or in motion, is a creative task and the creator must select visuals, color, style, type size, background and other aspects every time a new graphic is developed. He will have absorbed certain requirements and then visualized the way the selected elements should be graphically presented. The creative task of developing the graphic takes place in the person's mind with lighting speed and with no apparent effort. The finished product stands complete in his mind's eye; a clear translation of an idea into a graphic which is, unfortunately, invisible to everyone else.

Between creative moment and the completed "hard copy" of the graphic, the physical manipulation of tools tends to interfere with the development process, the more familiar and comfortable the tool, the less the interference. To minimize this interference, all graphics systems make some attempt to provide a "user friendly" man-machine interface, but in many cases they are friendly only to their oldest and most familiar acquaintances. The 3rd and 4th generation graphics manipulation devices are trying to walk the fine line between 'user friendly', often found to be so simple that the first time user gets instant gratification (but is soon frustrated by its limitations) and user specific, almost impossible for the uninitiated to comprehend but blindingly fast for a regular operator.

Ergonomic interface design, possibly coupled with some standardization of interface presentation, will continue to exercise and stimulate user to vendor dialog for many years to come.

Future Directions

The price/performance ratio in the area of imaging is improving simultaneously with a decrease in the cost of computer memory and processing power, to which it is closely linked. With more and more spillover of knowledge between the two disciplines, and the coming availability of new storage and retrieval techniques including Compact Disk-Read Only Memory (CD-ROM) and Direct Read After Write (DRAW) optical disks, the user will have access to some very sophisticated graphic generation and storage tools. Inexpensive storage and relatively fast retrieval will greatly enhance graphic utilization for library, archival and business functions.

It is clear today that many videographic systems are made up of multiple expensive digital devices that are interconnected via the analog domain. With increased demands for improved cost and performance objectives vendors designs will follow three main paths:

- Significantly cheaper "PC-based" equivalent products.
- Dedicated hardware based on high speed customized VLSI circuits.
- Integrated digital systems with total control of all video acquisition, manipulation and storage in the digital domain.

Lower-cost, "PC-based" manipulation and paint systems will soon be offering many of the capabilities at near the quality seen today in high-end broadcast units. The advent of PCs as the manipulation device in videographics is only beginning, but already there are paint systems, character generators and Digital Video Effects (DVE) devices appearing based on IBM PC XT and AT hardware. Potentially the 80386 microcomputer-based new generation PCs will achieve speed and memory manipulation increments of a high order.

Costs associated with VSLI development as well as a steep learning curve associated with their efficient application have slowed progress, but the industry is now ready to apply this technology effectively. Similarly the advent of high capacity disk storage and the DTTR have motivated the first fully integrated all digital concepts, where analog video is but a window to look at the pristine flow of digital images.

Standardization. Standards committees are close to finalization of standards related to both component and digitized video. This means that handling and manipulating component video in a localized area will become easier as manufacturers bring out or modify existing products to those standards. It also means the video world will become more adept at handling digitized video, which will generate a move to more 'computer-like' techniques of transmission, storage, error correction, etc.

With this increased ability to handle digitized video, Local Area Networks will become more popular in the video area, driven at first by the need to interconnect the new digital videotape recorders with digital still stores and manipulation devices. Local area networking (LAN) of video-based workstations having access to significant amounts of video storage and computer power will reduce the "cost per seat" for video artists, as well as encouraging interaction between talents during the creation of visual materials.

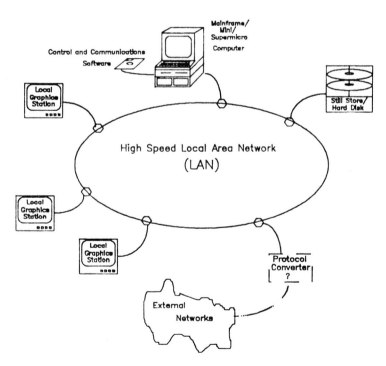

Video Local Area Network

Fig. 3.

Conclusion

It would seem that the current state of videographics forms an admirable base for creative and technological advances. Hardware innovation alone will change completely our approaches to information manipulation and storage, but it will be creative ability and demand that define the orientation and level of success of this growth industry.

Peter C. Lowten manages a consulting group in Mountain View, CA. After some years in R&D and video operations at the British Broadcasting Corp., he assumed various appointments with International Video Corp., Sony Broadcast and Compression Labs in the general disciplines of product management and marketing. Recent activities include a series of seminars for the Defense Communications Agency on Videographics and Videoconferencing and a project for the European Economic Community in Brussels. He is active in various technical societies and is a Charter Member of the Stanford Area Rotary Club. Mr. Lowten holds an Honors Degree in Engineering from Oxford University.

The Video Computer: Image Computing in the Studio

Alvy Ray Smith
Pixar
San Rafael, California

ABSTRACT

A general-purpose *video computer* is proposed which combines many studio or post-production functions, now available only in separate pieces of equipment, and extends their functionality. The ideal machine is described and the current state of the idea is given. The restrictions of realtime and broadcast day are compared. The conclusion is that video computers are already a viable idea within the broadcast-day-turnaround criterion and that the hardware exists as so-called *image computers,* general-purpose digital computers for the class of computations on images. Consequently software houses could immediately begin preparing applications on video computers for the broadcast video market.

INTRODUCTION

A look around a modern broadcast video studio or post-production house reveals a collection of machines made possible by the ongoing digital technology revolution:

- A digital transformation box (zoom, rotate, perspective)
- A digital effects box (page turn, flag wave, etc.)
- A digital wipe machine (perhaps inside a switcher)
- A digital paint program
- A digital font generation box
- A digital standards conforming box
- A digital matting box
- A digital polygon rendering box
- And others

In general, a different piece of hardware is purchased - at considerable expense - for each function. Each of these boxes may be thought of as a *special-purpose digital computer*.

An old notion in computing, now widespread in the general population, is that of the *general-purpose digital computer*. This is, of course, one machine that may be programmed to simulate all special-purpose machines. Since these fully programmable machines are commonly available, and since all the functions listed above have in fact been programmed on them, why are video studios still purchasing expensive special-purpose computers? The answer has several parts:

Computing Bandwidth

The broadcast video market has high computing bandwidth demands. A frequently stated requirement is that a function be computed in realtime, which for NTSC video means a new picture must be computed every thirtieth of a second. Assume a video frame nominally contains 512 by 512 pixels. This is near enough to 640 by 480, 512 by 488, and other popular video raster representations for estimates. So a video frame contains .25 megapixels. Further assume that each pixel is represented by three bytes - one each for RGB. Then a frame consists of .75 megabytes nominally. Furthermore, assume that to compute an interesting byte requires execution of 100-1000 computer instructions. Then the computing bandwidth required for realtime video applications is 75-750 million instructions per thirtieth of a second, or 225-2250 million instructions per second (mips). The only general-purpose digital computers which can approach even the low end of this range are the million-dollar supercomputers. Broadly affordable general-purpose machines are in the .1-10 mips range. So special-purpose machines have been required in video thus far. The video computer described below is a solution between these two extremes powerwise, but at the lower end pricewise.

Programming Talent

A typical studio lacks programmers. This might be changing, but it is probably safe to say that programmers capable of implementing, say, polygon rendering, are not generally to be found in the video studio or post-production house. Special-purpose boxes obviate programmers, of course, their ''programs'' being hardwired in. Any programmable box would most likely be accepted by this market only if applications software packages were readily available for the desired functions. Advanced houses could add functionality with their own programmers but the typical houses would simply buy functions as they could afford them, and never hire programmers.

Algorithm Availability

Some of the algorithms necessary for implementing the desired functions have not been generally available, so programmers would have nothing to program anyway. These include some of the most sophisticated functions - such as full-blown polygon and patch rendering, with antialiasing, texture, bump, and environment mapping, motion blur, follow focus, depth of field, shadows, reflections, refractions, etc. Again, packages marketed by software houses specializing in video computing programs would be the natural solution.

THE VIDEO COMPUTER - A DEFINITION

Another old idea in computing is that speed can be obtained by trading off against generality. The special-purpose boxes listed above are an extreme application of this idea. The ability to implement a vast number of functions at 1 mips is traded for implementing just one function at 100 mips. The *video computer* is neither fully general-purpose nor fully special-purpose. It is a general-purpose machine for video computations. As such, it is an instance of the larger idea of *image computer* which is a general-purpose computer for image computations, including all video computations as a subset.

How do image computations - hence video computations - differ from ordinary computations? First of all, they *are* ordinary computations in the sense that they can - and have been - implemented (but slowly) on ordinary general-purpose computers. Second, they are distinguished from all other computations in several obvious regards:

- The natural data type is the pixel - not the byte or word
- The pixels are organized in a rectangular array, or raster
- Color is a natural attribute of each pixel

The interesting point is that these seemingly simple specializations make possible - when traded off against full generality (e.g., income tax computation) - a cost-effective image computer, or video computer, capable of realizing all the desired functions very fast.

The Ideal Video Computer

The ideal machine would implement all desired functions in the video market in realtime and would be broadly affordable - comparable in price to just one of today's special-purpose boxes at say $50,000 to $100,000 at introduction. This really is the key statement of the ideal. It could be amplifed to include ease of programming, ease of interfacing, ease of use, etc., but these will follow the essential requirements of speed and cost. The pressures of the marketplace would combine with the continuing decrease in cost per mips to drive the price down from the introduction price, while maintaining functionality.

A Broadened Notion of the Ideal

The ideal video computer - as just described - cannot yet exist. The realtime restriction is the problem in 1987. Closer scrutiny of this "requirement" yields interesting results. Observation of the actual use of so-called realtime machines reveals that it is often the case that their use is slowed by preparation time or multiple passes.

For example, a digital transformation box might be used to rotate a logo while flying it along a curved arc. The box could indeed perform this function in realtime - i.e., as fast as the controlling knob, joystick, trackball, or other device were moved. But the reality of human control is that we are jerky, change our minds, get distracted, etc. and therefore have to practice realtime moves. Or, if the computer itself is used to make the movement - say, under spline control - then several paths will probably be tried before one is selected as desirable.

Another cause of slowdown can be the fact that to get a desired effect, a special-purpose box might be used in several passes and then the results composited in yet another pass or series of passes. For example, suppose a scene is mapped onto the eight surfaces of a double pyramid which spins and then unfolds so that the eight triangles spin off screen along eight separate paths. One possible way to accomplish this is to choreograph eight separate paths, one for each triangle and its matte, then composite them in appropriate juxtaposition and front-to-back ordering using the mattes for correct combination.

None of this is to deny that realtime is important. It is rather to point out that perhaps a more useful time measure in many instances is the *broadcast day*. The idea is that a machine may be sufficiently fast if an effect or other result can be obtained the very day it is conceived so that it can be aired that evening.

Thus, in the multi-pass example above, it would probably be sufficient to have a computer which could compute and composite all eight triangles in each frame rapidly - but not in realtime - and do it all in one pass and in, say, half an hour for a 30-second shot, including recording. Or eighty triangles.

Where Is the Concept Today?

An image computer already exists which meets the broadcast day requirement for those functions already programmed. This is the Pixar Image Computer which computes at 40 mips. It is in fact a video computer since it generates broadcast quality NTSC or PAL compatible RGB as standard outputs. It also falls within the introductory price range mentioned above.

Since some kind of realtime feedback is very useful for interactive design, a useful frontend for today's video computer is a realtime vector machine which can be used for realtime preview of effects using outlines to represent the final forms.

At this writing, it is the software applications programs which are still largely missing, but early examples of software houses already contributing to this marketplace are Alias Research Incorporated, Symbolics Inc., TASC/WSI, and Wavefront Technologies, providing applications for the Pixar Image Computer.

A GLIMPSE AT THE NEAR FUTURE

Consider the following image computations:

• A first-rate rendering of a complex 3-D object - a car, a spacecraft, a tree, a landscape, a face - with multiple light sources, full antialiasing, transparency, full hidden surface removal, bump mapping, texture mapping, and environment mapping - is generated at supercomputer speeds. An entire spline-controlled animation is generated easily within a broadcast day. The object contains 100,000 polygons - or 1,000,000 polygons.

• A weather report shows roiling 3-D clouds casting shadows as they move across a bas-relief representation of the northeastern United States. A hurricane forms off the coast.

• A production assistant wraps a frame onto a disk and spins it offscreen right, revealing the scene behind, or twists a picture about its midline with the twist proceeding from left to right in an accelerating motion to ''throw'' the audience's attention that direction. There are no jaggies anywhere. Both effects are unavailable on special-purpose boxes. All standard rotates, scales, and perspectives are of course available too, as are all standard wipes.

• The full generality of a so-called ''paint program'' is invoked by the staff artist for a quick graphic or an exquisite painting for a background.

• A reactor accident occurs in the Ukraine. A quick call to a national earth resources database gets the studio a satellite photo of the area taken since the accident. The area of interest is magnified, the edges are sharpened, contrast is enhanced - all with standard image processing routines. Text is added and the result goes on the air.

• A difficult blue-screen matte is pulled using techniques not available without digital computations localized to the pixel level.

The important point is that: **One box - a video computer, or image computer - does all of these.** An effects box, paint computer, matting box, wipe box, transformation box, image processor, polygon-rendering computer are folded into one - and still other functions can be imagined and performed.

IMPLICATIONS

A partial listing of the implications explicit and implicit in the video computer concept is:
• One box instead of many
• Repeatable procedures - change one small thing but repeat everything else
• Graphics and effects production brought in-house
• Broadcast-day turnaround
• A natural growth and upgrade path - add functionality as needed
• Opportunities for vendors to provide advanced software applications
• Opportunities for vendors and customers to customize a look

Once general-purpose computation has entered the studio full-force, then all the powerful concepts which have evolved for the office and university become applicable to the studio: Operating systems, database managers, networks, communications standards, common interfaces, archiving, expert systems, etc.

Dr. Alvy Ray Smith co-founder and vice-president of Pixar, a company specializing in digital visualization products and services. He has founded during his career, with Dr. Edwin Catmull, three centers of computer graphics excellence: the Computer Graphics Laboratory at the New York Institute of Technology in the mid 70s, the Computer Division of Lucasfilm Ltd. in the early 80s, and Pixar, a year ago. Alvy has published many technical papers in computer graphics, created or directed some of its best-known pictures and animations, and managed some of the most innovative people in the field, in hardware and software. He received his Ph.D. from Stanford in 1970.

Variable-Resolution Rendering System Extends Television Animation Graphics to Film and Print Media

Dr. Philip Lucht
BTS Broadcast Television Systems, Inc.
Salt Lake City, Utah

ABSTRACT

Problems encountered in transferring computer–generated television–resolution images to film and print media are briefly reviewed, including the effects of anti–aliasing. It is argued that, in order to fully utilize the spatial resolution these media provide, pictures must be completely re-rendered directly at high resolution. This introduces a new set of problems which are briefly summarized. When previewing requirements are considered, as well as the final product, rendering systems allowing completely variable resolution seem most appropriate. Features of such a system currently under development at BTS Broadcast Television Systems, Inc. are described.

INTRODUCTION

The last five years have seen a rapid growth in the use of computer animation graphics for television, and a corresponding growth in the number of companies offering turnkey products designed to produce these graphic effects. Such products now range from PC-based systems which are relatively low cost but slow, to hardware-intensive high-end systems which are more expensive but have higher production throughput.

All these systems have a common set of goals: to allow their users to set up the geometry and surface attributes of 2D and 3D graphic objects (Modeling), to describe how these objects, along with light sources and simulated camera and environment parameters, should move or change in time (Animation), to construct the individual frames of one or more animated image layers (Rendering), and finally to combine these image layers into a final product, with the possible admixture of live video information (Compositing).

Typically, the animation frames are rendered into a television-resolution digital frame buffer, and are subsequently recorded frame-at-a-time onto video tape or video disk. The per-frame render time can vary from a fraction of a second to many hours depending on the image complexity, the power of the rendering hardware, and the nature of the rendering software.

Although some systems use run-length encoding in their frame buffers, most store individual pixels. The number of pixels per scan line is usually in the range 512 to 1024, while the number of visible scan lines is forced by the NTSC specification to be about 485 (585 for PAL). These numbers are characteristic of "television resolution" and, with the use of reasonable filtering algorithms, are generally considered sufficient for making a good quality image, considering the nature of the medium.

To some extent, the number of effective vertical and horizontal samples can be increased by the use of supersampling (computing the picture at higher than displayed resolution) and subsequent spatial filtering to remove the high-frequency information from the picture, an operation commonly referred to as anti-aliasing. The general effect is to replace highly defined but jagged-looking edges with smooth but somewhat blurred edges.

As a concrete example of the notions mentioned so far, we quote the characteristics of a product developed by the Robert Bosch Corporation [1]. The FGS-4000 is a relatively high-end television graphics system in terms of the price/performance spectrum alluded to earlier. It contains digital frame buffers which have 768 visible pixels per scan line. Pictures are supersampled at four times screen resolution and are then filtered down symmetrically in both horizontal and vertical directions using a 7 x 7 overlapping Gaussian weighted filter mask. The resulting picture quality is generally considered to be quite good when viewed on either an RGB or composite monitor.

THE HIGH RESOLUTION PROBLEM

Although pictures produced on the FGS-4000 and similar systems may be excellent when viewed on a television monitor, attempts to transfer these same pictures to film or print media frequently result in a noticeable deterioration of the apparent picture quality. The reason of course is that there is a sudden dramatic increase in the quality of the film or print medium, but no increase in the quality of the video picture information. Whereas typical four-color printing plate technology can resolve 300 to 1800 dots per inch, and film has a resolving power of 50 to 600 lines per millimeter, the vertical television-resolution picture information has a density of about 500 lines per foot as seen on a typical television monitor.

Commercially available film printers attempt to deal with this resolution conversion problem in various ways. The lower-cost analog instruments contain a CRT which connects directly to the analog RGB signals from the rendering system frame buffer, and provide a Polaroid-style camera to directly photograph the monitor face using sequential color filters. A microprocessor controller might attempt to defocus or dither the scan lines slightly in an attempt to fill the interline gaps.

In a digital film printer, the numerical pixel values which describe the rendered frame are digitally transmitted from the rendering device to the film printer, either directly through a hardware interface, or indirectly by means of a standard half-inch magnetic computer tape. The digital method avoids the signal quality loss associated with the analog circuitry of the rendering device, which is only designed to operate at normal television bandwidths. Eventually, the digital information is converted to analog within the film printer system, where it is applied to a very high resolution CRT. Such digital film printers are capable of resolving up to 8096 samples in both directions, and for that reason the systems are relatively expensive.

How then is one to stretch the 485 scan lines of digital television resolution information to the 2K, 4K or 8K scan lines (K=1024) required by a digital film printer? Methods range from simple pixel replication and (much better) linear interpolation, to more sophisticated proprietary interpolation and resampling techniques. These processes are normally performed by a host computer attached to the digital film printer.

The success of these digital post-processing methods depends on the picture content. Soft features such as clouds and airbrushed edges come out well. However, features which contain information whose spatial frequency exceeds the resolution of the rendering device -- such as fine lines and text -- will never look right on film regardless of the resolution of the film printing device. In order for fine detail to exist in any reasonable sense (no jaggies) in the television-resolution frame buffer, the renderer had to use anti-aliasing techniques to blur the edges. Once blurred, these edges can never be restored to film-resolution sharpness.

It is simply a matter of lack of information: the extremely high resolution feature coordinates from the original graphics data base objects, ideally 64-bit floating-point numbers with an effective resolution of about 1 part in 10^{16}, have been heavily low-pass filtered so that only 10^3 or so samples are left along an axis of the television resolution picture.

HIGH RESOLUTION RENDERING

In order to fully utilize the high spatial resolution of film and print media, there is only one solution to the problem described above. The television resolution picture must not be used at all. The picture must be re-rendered all over again from its original data base at the desired film printer resolution. The rendering software should be capable of rendering at arbitrary resolution and aspect ratio. For example, a 35 mm slide with a 3:2 aspect ratio might be rendered with 1365 scan lines of 2048 pixels each, whereas a square magazine graphic may require 4096 x 4096 pixels.

Even at these higher resolutions, the rendering algorithm should continue to use supersampling/filtering techniques to remove aliasing artifacts which can affect the shading, texturing, and edge quality of the picture.

The benefits of high-resolution rendering become obvious when one views an 8 x 10 print. Figures 1 and 2 show portions of two such prints of a test pattern done on a Genigraphics 8770 Masterpiece film printer. The pattern contains many near-horizontal and near-vertical edges to accentuate edge aliasing problems. Figure 1 was rendered at 512 x 512 and pixel-replicated up to 4K x 4K, while Figure 2 was rendered directly at 4K x 4K. Both frames were supersampled at four times resolution and were then filtered down using the FGS-4000 filter described above. If the difference between these two figures is not immediately visible to the reader, it is due to the nature of the reproduction technique used to generate this paper.

Although the benefits of high-resolution rendering are attractive, the cost side must also be taken into account. While a television-resolution picture contains about a third of a million pixels, a 4K x 4K picture contains over 16 million pixels, a factor of 50 increase. To the extent that rendering time is proportional to the number of pixels rendered -- which is approximately true -- a 10 minute television-resolution frame render becomes an 8 hour task at 4K x 4K.

Associated with this increase in the pixel count are assorted data storage and transmission problems. At one byte per color component, a single 4K x 4K picture requires 50 MB (megabytes) of digital storage. A standard 2400-foot 9-track computer tape holds a mere 44 MB at 1600 BPI density and 171 MB at 6250 BPI (unformatted). Even with write-once optical disk stores, disk space is quickly consumed by pictures of such great size. And one does not even want to consider the time it would take to transfer such a picture over a serial line. (One should not even *think* of doing such a thing.)

Pictures can usually be compressed using run-length and other encoding schemes, but the compression factor decreases as picture complexity increases. On the other hand, the use of more than 8 bits per color component and the addition of a matte or "alpha" channel drives up the storage requirements even further.

Another operational difficulty with high-resolution rendering is that there is no cheap way to quickly preview the entire picture exactly as it will appear on the output medium. This is a problem of both resolution and color.

One can render the entire picture at television resolution and view it on a standard RGB monitor, or one can render a small television-resolution window of the full picture to see at least some pixels at their true resolution. Special purpose high-resolution frame-buffer/monitor systems having perhaps 2K x 2K resolution can be used, but cost is again a problem. A "standard" HDTV system (1035 visible scan lines of 1920 pixels each) provides an interesting intermediate possibility.

In the case of an animation where many frames are needed for a preview or motion test, it may be advantageous to render images at *below* television resolution to reduce rendering time for a fast raster motion test. Such frames are then pixel-replicated and displayed on a standard TV monitor.

As far as the color accuracy provided by any CRT previewing system is concerned -- viz-a-viz the final medium -- we can only say that it is matter of some controversy, and that many people are less than enthusiastic about luminous screens and calibration knobs.

Beyond the CRT systems there are various hard-copy devices available to fill the gap between television resolution and the final film or color separation plates. As the color and spatial resolution requirements on such devices are increased, costs seem to rise accordingly.

The varying requirements of film and print output media, as well as the intermediate media used for previewing, serve to highlight the need for a *variable* resolution rendering system. In the following section we shall describe some of the features of such a system currently in development at BTS.

BTS VARIABLE RESOLUTION RENDERING SYSTEM

BTS Broadcast Television Systems Inc., a joint company of Bosch and Philips (formerly Robert Bosch Corporation, Video Equipment Division), is currently developing a variable resolution rendering system to meet the general requirements described above.

The system is at present a purely software system written in the C language for use on a Sun Microsystems 3/110 UNIX workstation, and is intended to render animation frames that are built on the FGS-4000 Graphics System. Object and animation files are transmitted to the Sun from the FGS on an Ethernet link. Sun-rendered pixel files are then placed on the Sun's high density quarter-inch tape cartridge (60 MB per cartridge), and are later translated as needed to specific film printer formats and are placed on standard half-inch computer tape. Rendering resolution is continuously variable in both directions independently up to 32K x 32K pixels.

Prior to rendering, a wire-frame view is presented on the Sun screen. The animation can be moved in time, and desired frames selected for rendering. At this wire-frame previewing stage, the picture can also be cropped and shifted to match the aspect ratio of the intended final output medium.

The Sun 3/110 workstation contains an 8-bit 1152 x 900 greyscale frame buffer and monitor, so that one can preview a high-resolution frame by rendering it at any resolution smaller than 1152 x 900. Each of the RGB (or CMYK) color separations can then be previewed as a 256-level greyscale image. Frames which are preview-rendered at FGS-4000 resolution (768 x 485) can be sent back over Ethernet to the FGS for viewing in full 24-bit color (16 million colors).

Internally, the variable-resolution renderer maintains 16 bits per color component, allowing for a theoretical 280 trillion colors. Some film printers currently utilize up to 12 bits per component.

As each frame is rendered in RGB, a perfectly registered 16-bit soft matte is rendered as well. This so-called "alpha channel" can then be used as a transparency/key signal enabling rendered image layers to be arbitrarily composited together after they are independently rendered [2]. The compositing process runs much faster than rendering, so a correction in one layer can be made by re-rendering only that layer and then recompositing all layers together again.

Within each image layer, frames can be rendered using back-to-front priority, or alternatively, a 32-bit depth buffer can be activated to allow for object interpenetration. In this mode, opaque objects are rendered front-to-back, and partially transparent objects are properly handled by doing a depth-sorted microcomposite at supersampled resolution. Supersampling itself is selectable at 1X, 2X or 4X with an appropriate overlapping Gaussian filter to give a high quality anti-aliased image.

As on the FGS-4000, curves are dynamically expanded based on their screen size to provide a characteristic smooth edge. A smooth-shading model is provided to emulate FGS-4000 shading, or users can provide their own shading equations in C if they so desire. The full FGS complement of sixteen light sources is provided. In addition, it may be possible for users to supply their own texturing functions which can be applied to smooth-shaded objects, or from which such objects can be sculpted. Software hooks are provided for the rendering of externally introduced objects constructed of bicubic surface patches.

All geometric and shading computations are done in floating–point math using the 68881 coprocessor chip included in the Sun 68020–based workstation. In this way, dynamic range problems inherent in integer–math systems are avoided. It is interesting to note that a large fraction of rendering time is connected with floating point operations, and that certain state–of–the–art VLSI floating point chips and chipsets –– when connected to bit–slice processing elements –– can run approximately 100 times faster than the 68881.

CONCLUSION

In this brief report we have attempted to describe some of the problems involved in extending television–resolution computer graphics to the higher quality film and print media. For lack of space and time, many topics have gone quietly unmentioned, such as the question of converting live video to high resolution, or of loading in real–world pictures using laser scanners.

We have argued the need for a variable resolution rendering system, and have described the characteristics of such a system currently under development at BTS. A skeptic might argue that any rendering system is automatically "variable resolution" by simply changing some "software parameters." We have found that this is quite definitely not the case.

Many of the problems encountered in high resolution rendering are quite severe, especially when animations with many frames are involved. As the discussion has suggested, high–resolution rendering and high–powered parallel hardware will likely have a fortuitous rendezvous as technologies continue to progress.

Footnotes:

1.   J.A. Briggs, "Electronic Graphics in Television – The Next Step," SMPTE Journal, Vol. 92, No. 9, Sept. 1983, pp. 912–917.

2.   T. Porter and T. Duff, "Compositing Digital Images," SIGGRAPH '84 Conference Proceedings, ACM (July 1984), pp. 253–259.

Dr. Lucht has been associated with the Robert Bosch Corp. in various engineering and management capacities since 1980. He contributed to the architecture and hardware design of the Bosch FGS-4000 Graphics System, and is currently engaged in new product design for BTS Broadcast Television Systems, Inc., a joint company of Bosch and Philips.

Fig. 1. An Anti-Aliased Test Pattern Rendered at **512 × 512**
and Pixel replicated to **4K × 4K**.

Fig. 2. The Same Test Pattern rendered directly at **4K × 4K**.

R-DAT Format Overview

P. A. Dare and R. Katsumi
Sony Communications Products Co.
Teaneck, New Jersey

INTRODUCTION

It is now some 35 years since $\frac{1}{4}$ inch analog audio tape recording was introduced to the broadcasting and television industry. The $\frac{1}{4}$ inch analog type format, whether it be in its full track or $\frac{1}{2}$ track modes, has served the industry well, in fact; there are very few standards that have survived 35 years in the audio recording industry. That standard is now challenged by the R-DAT proposal; R-DAT being an acronym for Rotating Digital Audio Tape. In addition to potentially changing the industry's attitude to the $\frac{1}{4}$ inch analog tape format, the R-DAT format potentially challenges the tried and trusted NAB cassette. I would like to describe to you the current status of the R-DAT standard. While the 1983 worldwide DAT conference did not discuss the professional R-DAT standard there is little doubt that the professional standard will track very closely the consumer format that has been proposed.

R-DAT HISTORY

R-DAT dates back to 1983. In June of that year Sony Corporation unveiled an entirely new digital audio cassette concept based upon the unique rotary helical scanning technology. This recording method which is well known to the television industry was previously only found in video recorders. It offered the advantages of miniature size, simplicity, and extremely long recording time on a very small cassette, at a much lower production cost than had been previously estimated for any stationary head design. The R-DAT format was formally finalized this last spring [1986]. Sony showed the first working prototype in the United States at NAB in April of 1986. More recently at the Consumer Electronics Show in Las Vegas a large number of manufacturers showed advanced prototype models of the consumer format machines. For the purpose of being complete, I have no intention at this time of covering the status or condition of the future S-DAT which employs a fixed recording head.

R-DAT FORMAT

The R-DAT format is a digital format. The digitalized audio signal is recorded on magnetic tape in a format as shown in Figure 1. The standard track width is 13.59 micrometers, the track length 23.5 mm, and the minimum recorded wave length is 0.67 micrometers. For purposes of comparison, in the video world a minimum recorded wave length of 0.9 micrometers is considered extremely small. The packing density of R-DAT is 114 Mb/s per square inch, all of this being achieved with a Linear tape speed of 8.1 mm per second. For comparison purposes, the tape speed of the analog compact cassette is 47.6 mm per second. As part of the R-DAT format (see Table 1), it is possible to use the standard 48 KHZ 16 bit sampling system or alternatively 44.1 KHZ/16 bit or 32KHZ/12-16 Bit sampling. The 44.1 KHZ system is only available in the professional R-DAT machines. The consumer R-DAT machines are restricted to 48 KHZ record/play, and 44.1 playback only. As shown in figure 2, the scanner has 2 heads; the record playback heads are in contact with the tape 50% of the time; and for the other 50% of the time no contact with the tape is made. The use of such a discontinuous signal has the following advantages. The 90° wrap angle means only a short length of tape is in contact with the drum at any time, this reduces tape damage and allows high speed search while the tape is in contact with the drum. The search speed of the professional R-DAT is in the order of 300 times normal play speed. Low tape tension is used which ensures long head life. As an alternative to the 2 head construction, 4 heads may be used. The heads can be separated by 90% to allow for simultaneous playback. One implementation of the scanner configuration uses a 30 mm diameter scanner and is rotated 2000 RPM; as a result it has a resistance to external disturbances somewhat like a gyroscope. Under these conditions the 2.46 Mb/s signal to be recorded, which includes audio, as well as many other types of data, is compressed by a factor of 3 and processed at 7.5 Mb/s per second. This enables the signal to be recorded discontinuously. To overcome the well-known low frequency problems of coupling transformers, the channel code has an 8 to 10 conversion in which the 8 Bit signal is converted to a 10 Bit signal. This channel coding also has the advantage of reducing the range of wave lengths to be recorded. The maximum wave length is 4 times longer than the minimum wave length. It also makes over-writing possible which eliminates a separate erase head. Over-writing is easily accomplished when short wave length signals are to be erased. In addition to the two recording heads use is made of a plus or minus 20 degrees azimuth recording system without guard bands. To implement this recording system we are using playback heads that are about 50% wider than the tracks that remain on the tape. As will be explained later this will enable automatic track following.

R DAT ERROR CORRECTION

As with any digital recording format consideration must be given to the error detection and correction scheme. (See Figure 3 and Table 2.) In addition to having to correct the digital audio data, correction and detection is also provided for the sub codes, ID codes, and other auxiliary data. The term "burst error" refers to drop outs caused by dust, scratches, and clogging of a head. Random errors may be caused by cross talk from an adjacent track, traces of an in perfectly erased or over-written signal, or mechanical instability. To cope with these errors, the R-DAT format is provided with double Reed Soloman codes which are very effective for error detection, correction and interpolation, the typical ECC overhead being in the 40% range. The signal is rearranged as interleave blocks to make playback possible even if one head is clogged. The blocks are recorded straddled across two tracks. The error correction code interleave length is one track while audio data interleave length is 2 tracks. The error correction code format of R-DAT is the same as that of a product code. Each error correction code $(C1/C2)_8$ uses the Reed Soloman code over a Galois field GF (2^8) as shown in Figure 4. The product code of D=5 and D=7 gives this format a powerful correction capability for random errors. This format also shows excellent performance for burst errors by means of the interleaving of data. For a burst error the correction data length is 792 symbols and the concealment length is 2664 symbols. The purpose of the error control system is the prevention of click noise with any playback data. There are a few cases where there is a possibility of unsatisfactory Cl/C2 decoding. The error control system then needs a feed forward control to analyze detected data quality. The controls, the decoder and the resulting action are summarized in Table 3. In conjunction with the error correction and detection system, it is also interesting to consider what happens during variable speed playback. The data from the left and right channels are disposed diagonally across two tracks as illustrated in Figure 5. Sound reproduction of good quality is possible even for variable speed playback with the use of average value interpolation. The format of this interleaving makes it possible to generate necessary data even if only one of the two heads can read out the data as is the case of variable speed playback where the head jumps over the track. No ECC interleave error will occur because of its one track interleave length while appreciable interleave errors exist in the audio data. The error rates of the data from the two heads are compared in variable speed playback. The data with the smaller error rate is used to reconstruct sound signals and the other data is discarded. The missing data necessary for decoding is obtained from the chosen data by a process called compulsory interpolation. In addition to reproducing audio during the variable speed playback mode, R-DAT is capable of high speed music search. The R-DAT system allows high speed music search up to 300 times its normal speed. In addition to monitoring the linear speed of the tape, the rotational speed of the drum is controlled between 1,000 and 3,000 rpm to assist in providing the sound quality during the 300 times search mode. It is necessary to control the variation of relative speed of the tape to head, the off tape data must be kept within the PLL lock range. The relative speed is given by:

$$vR = \sqrt{(vD \cos \theta o - Nvt)^2 + (vD \sin \theta o)^2}$$

where vD = Drum Speed
θo = Stationary angle
Nvt = Tape Speed

Using a 30mm scanner, scanner speeds of 1000-3000 RPM can be expected.

TRACK FORMAT

The audio data which has been subjected to the 8:10 channel code conversation, double Reed Soloman ECC coding and interleaving and other processing, also has a block format in which each block consists of 288 bits (see Table 4). A close look at the block format will show with the exception of the sync byte all block address interleaved contents can be checked by 8x3 = 24 bits. This is also useful for accurately identifying the start of tracks. The 256 bits of data consist of 32 symbols (one symbol is represented by 8 bits). One track contains 128 blocks consisting of these symbols (a total 4,096 symbols). Of these 1,184 symbols (C1 and C2 are included) are used for error correction which leaves 2,912 symbols for use as data. There is also a reserved area for sub codes which will be utilized later. The sub codes are mainly used for recording program numbers and time codes. The sub codes are in two locations above and below the tracks which makes them resistant to drop outs. The areas for ID codes associated with the PCM areas are used for recording various items in the same manner as compact discs, such as sampling frequency, channel number, quantization number, tape speed, copy protection, use, or non-use of emphasis, etc.

R-DAT ATF:

A unique automatic track finding signal is recorded along with the digital data. As shown in Figure 6 the ATF makes use of a pilot signal F1, Sync signal 1 F2, Sync 2 signal F3, and erase signal F4. When the head advances in the direction of the arrow, the presence of a ATF signal is detected by either F2 or F3 signals. The adjacent pilot signals F1 on both sides are then immediately compared and then a decision made whether the tracking is correct or not. The F1 signal components use a low range frequencies that are not affected by the azimuth setting so crosstalk can be picked up and detected from both sides. Since this AFT signal compares the cross talk using an analog method, the processing is different than that for other areas. The AFT area is clearly divided into two parts in the track format, so a small amount of track curvature does not result in tracking errors. This together with the use of wide heads has resulted in a system in which compatibility is easy to achieve. A tracking control head such as that used for the VTR is not required using this ATF system.

LSI FOR R-DAT

Currently in product development are 5 LSI's. These are being designed to realize the signal processing described in the previous sections (see Table 5). The LSI's perform the function of error correction and interpolation, modulator and demodulator, data control IC, Servo IC, auto track following system, FM amplification and equalization IC. The first 3 of the integrated circuits are C Mos, the remaining, Bipolar. The chips where designed to include:

1 An error control system to prevent click noise while playing back any data;

2 A wide margin for signal processing to compensate for mechanical misalignment;

<u>3</u> Versatility so that the chips can be used in high price and low price machines, and digital audio equipment other that R-DAT;

<u>4</u> Expandability of the functions of the R-DAT system. Figure 7 is a block diagram of an R-DAT system utilizing the LSI's that were just described.

DUPLICATION

High speed duplication of R-DAT while not a reality today is more than a possibility in the future. The magnetic contact printing technique is one in which the magnetic surfaces of a pre-recorded mother tape and that of a copy tape are put together in contact with each other (see Figure 8). A magnetic bias is applied to the contact area. The printing process can be called the ideal anhysteretic magnetizing process. Firstly, the mother tape and the copy tape are tightly squeezed together. Secondly, the printing drum rotates and drives a mother and copy tape. The tapes are pressed to the printing drum by air pressure. During the printing process the tension on the mother and the copy tapes around the printing drum is carefully regulated. As you can see from this figure, the 90 degree wrap angle that was described earlier as having advantages during high speed search, also has an advantage in the printing process. Certain limitations exist in the magnetic contact printing process. The coercivity of the mother tape has to be approximately 3 times that of the copy tape. The tape utilized in the R-DAT mastering is metal particle tape which permits the recording of very short wave lengths. Therefore, tapes suitable for the copying must have coercivity in the region of 700 oersteds and capability recording small wave lengths. It is interesting to note that the Barrium ferrite tapes currently under development have this capability. Barrium ferrite tapes can be produced with coercivity in the range of 700 oersteds with the capability of recording very small wave lengths. Perhaps the question may be asked within the magnetic contact printing process; why isn't the mother tape erased? The solution to this is to apply the bias magnetic field perpendicular to the surface of the tape in order to reduce the demagnitization effect (See Figure 9.) In addition, soft iron is used in the printing drum in order to increase the perpendicular component of the bias magnetic field. The other question that probably remains is how does the copy tape compare to the original (See Table 6). Electronically, the output of the shortest wavelength 0.67 micrometers is 3:5 db lower than that of the conventional copied tape. In reality this is not a problem as the recorders have sufficient head room to make up for this deficiency in output. Another characteristic of hi-speed printing is that the auto track following signal of approximately 100 KHZ has a very low output. Because of this, the automatic tracking following signal of the mother tape must be pre-emphasized. The R.E.B. rate of the printed tape is less than 1 in 10^{-3} and with this condition the reproduced signal is fully corrected. The envelope fluctuation produced by mechanical and tracking errors is less than 1.5 db peak-to-peak. The track linearity is less than 3 micrometers peak-to-peak. It is envisaged as time passes that high speed printers capable of printing 150 times the normal speed will be produced.

R-DAT CASSETTE SIZE

The cassette is a completely sealed structure and measures 73 mm x 54 mm x 10.5 mm weighs about 20 grams. See Figure 10.

R-DAT THE FUTURE

Semiconductor makers are experiencing rapidly decreasing costs. This promises to make possible many new applications for equipment incorporating digital technology. If a very small drum can be produced, an even smaller digital tape recorder could fit in the palm of a hand. By halving the tape speed and changing the sampling frequencies, the recording time could be doubled from 4 to 6 hours, or 4 channels could be provided. Furthermore, by arranging for automatic mode selection by means of sub-codes, the tape recorder could have many new applications including use as an inexpensive data recorder or as an adjunct to a audio recording console.

CONCLUSION

The R-DAT and its associated equipment lends itself ideally to applications in the broadcast, television, and film industry. It certainly is the first challenge in 35 years to the analog ¼ inch format and to the currently used NAB cassette. It offers great opportunities to the broadcaster in automation systems and perhaps more important it offers to the listener at home the possibility of extremely high quality sound without the expense of high priced digital audio tape recorders.

ORGANIZATION OF THE DAT CONFERENCE

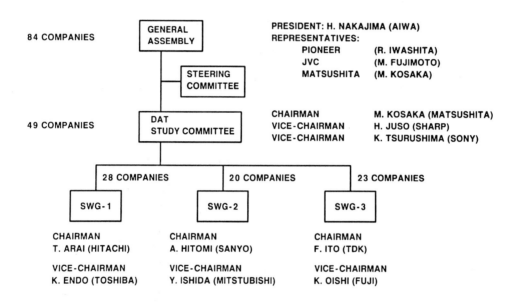

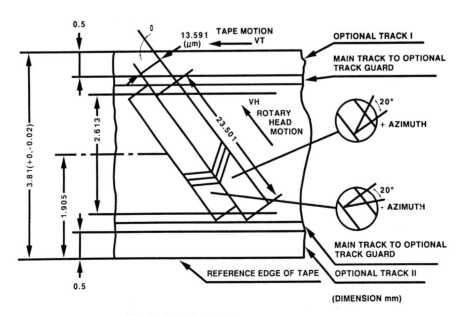

Fig. 1. R-DAT Track Pattern (Simplified).

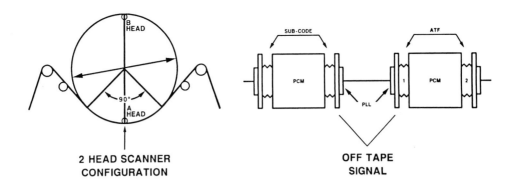

2 HEAD SCANNER
CONFIGURATION

OFF TAPE
SIGNAL

ORIGINAL DATA RATE ≃ 3.2Mb/s
COMPRESSED DATA RATE ≃ 7.5Mb/s

Fig. 2. R-DAT Scanner Configuration.

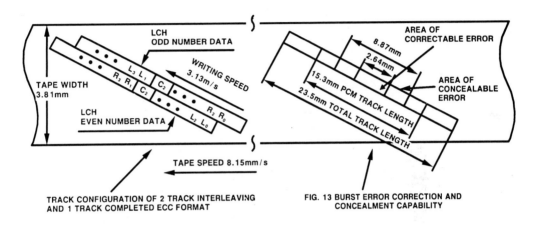

TRACK CONFIGURATION OF 2 TRACK INTERLEAVING
AND 1 TRACK COMPLETED ECC FORMAT

FIG. 13 BURST ERROR CORRECTION AND
CONCEALMENT CAPABILITY

Fig. 3. R-DAT Error Correction Capability.

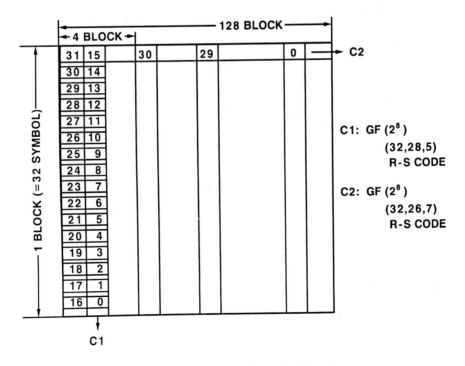

C1: GF (2^8)

(32,28,5)
R-S CODE

C2: GF (2^8)

(32,26,7)
R-S CODE

Fig. 4. R-DAT Error Correcting Format.

INTERLEAVE SCHEME

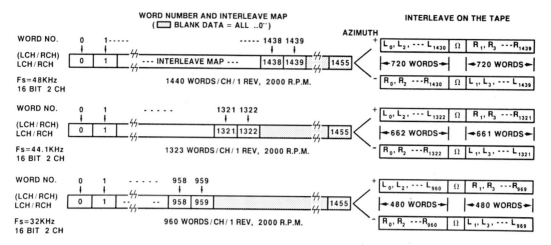

Fig. 5. R-DAT 3Fs Interleave Scheme (Fs=48, 44.1, .32KHz 16 BIT).

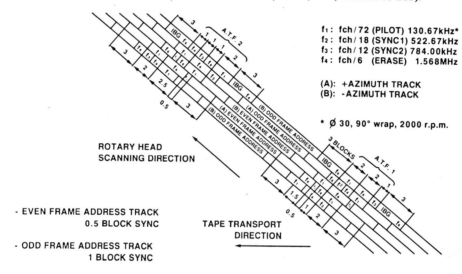

f_1 : fch/72 (PILOT) 130.67kHz*
f_2 : fch/18 (SYNC1) 522.67kHz
f_3 : fch/12 (SYNC2) 784.00kHz
f_4 : fch/6 (ERASE) 1.568MHz

(A): +AZIMUTH TRACK
(B): −AZIMUTH TRACK

* Ø 30, 90° wrap, 2000 r.p.m.

ROTARY HEAD
SCANNING DIRECTION

- EVEN FRAME ADDRESS TRACK
  0.5 BLOCK SYNC

- ODD FRAME ADDRESS TRACK
  1 BLOCK SYNC

TAPE TRANSPORT
DIRECTION

Fig. 6. R-DAT A.T.F. (Automatic Track Following).

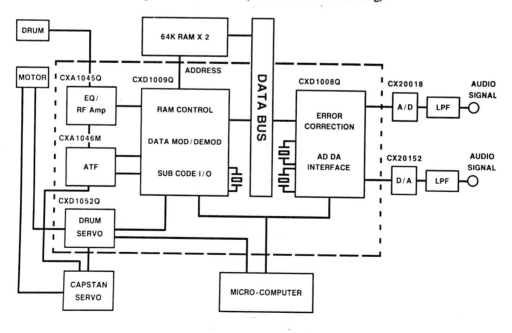

Fig. 7. R-DAT System.

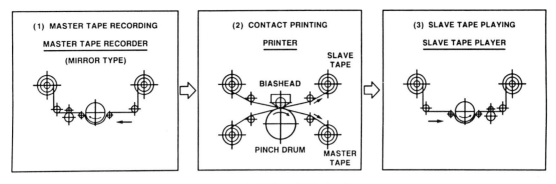

Fig. 8. R-DAT Hi Speed Printing Process.

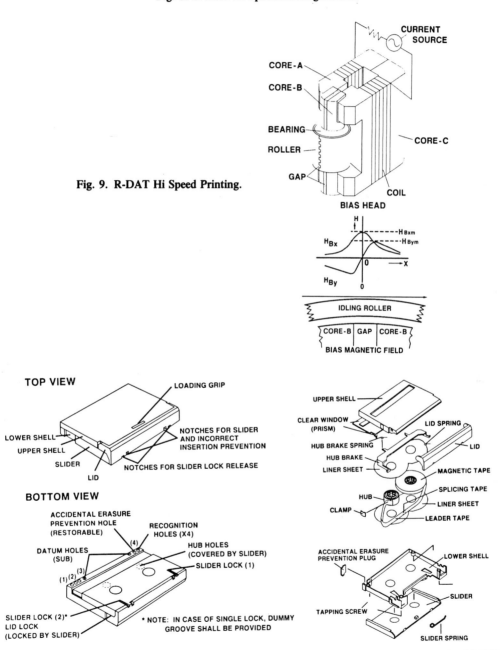

Fig. 9. R-DAT Hi Speed Printing.

DIMENSIONS 73 X 54 X 10.5mm

Fig. 10. R-DAT — Cassette.

		CH/unit					
NUMBER OF CHANNELS	(CH)	2	2	2	4	2
SAMPLING FREQUENCY	(kHz)	48	44.1	32		
QUANTIZATION	(BIT)	16	16	16		12
TAPE WIDTH	(MM)	3.81 (+0/ -0.02)				
TYPE OF TAPES		MP	OXIDE	MP		
TAPE THICKNESS	(μM)	13 ± 1				
TAPE SPEED	(MM/S)	8.15	12.225	8.15	8.15	4.075
TRACK PITCH	(μM)	13.59	20.41	13.59		
TRACK ANGLE (TAPE RUNS)	(DEG)	6°22'59.5"				
RECORDING TIME	(MIN)	120	80	120	120	240
HEAD GAP AZIMUTH ANGLE	(DEG)	±20				
RECOMMENDED CYLINDER SPECIFICATIONS		φ 30, 90° WRAP 2000 RPM				1000 RPM
WRITING SPEED	(M/S)	3.133	3.129	3.133	3.133	1.567
MODULATION SCHEME		8 - 10				
RECORDING DENSITY		61kBPI				
ERROR DETECTION AND CORRECTION CODE		DOUBLY-ENCODED RSC $\begin{pmatrix} C_1 & 32, & 28 & 5 \\ C_2 & 32, & 26 & 7 \end{pmatrix}$				
REDUNDANCY	(%)	37.5	42.6	58.3	37.5	37.5
TRANSMISSION RATE	(MBPS)	2.46				1.23
SUB-CODING (CAPACITY)	(kBPS)	273.1				136.5
TRACKING SYSTEM		ATF				
DIMENSION OF THE CASSETTE	(MM)	73 X 54 X 10.5 (W X D X H)				

Table 1. Basic Specifications of R-DAT System

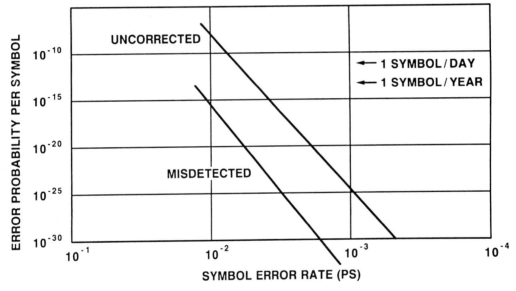

Table 2. R-DAT — E.C.C. Capability

	FACTORS	PROBLEMS	CONTROLS TO DECODER
A	1 TRACK INTERLEAVE LENGTH ECC AND 2 TRACK INTERLEAVE LENGTH AUDIO DATA	INTERLEAVE ERROR OF AUDIO DATA	IF FRAME ADDRESSES OF A AND B CHANNEL TRACKS NOT THE SAME, TRACK WITH THE LOWER ERROR RATE IS PLAYED BACK
B	2 BLOCK SIDE-BY-SIDE CONNECTION FOR THE INTERLEAVE LENGTH OF C1	PLAYBACK OVER SOME TRACKS OR ON UNERASERING AREA	C2 ERASURE CORRECTION PROHIBITED
C	REMARKABLE MIS-DETECTION WHEN SER ≥ 10^{-1}	PLAYBACK ON THE NO-RECORDED AREA	COMPULSORY MUTING

Table 3. R-DAT Feed — Forward Controls

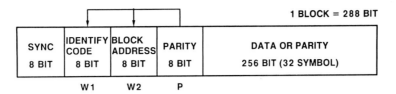

1 BLOCK = 288 BIT

SYNC 8 BIT	IDENTIFY CODE 8 BIT	BLOCK ADDRESS 8 BIT	PARITY 8 BIT	DATA OR PARITY 256 BIT (32 SYMBOL)

W1   W2   P

- PARITY: $P = W1 \oplus W2$ ($\oplus$: MOD 2)
- BLOCK ADDRESS: BLOCK ADDRESS FOR PCM DATA BLOCK OR SUB CODE BLOCK

  THE MSB IS IDENTIFY BIT FOR SUB CODE BLOCK OR PCM DATA BLOCK

MSB   LSB

| ..0" | * | * | * | * | * | * | * | : PCM BLOCK (BLOCK ADDRESS = 7 BIT)

| ..1" | | | | * | * | * | * | : SUB CODE BLOCK (BLOCK ADDRESS = 4 BIT)

Table 4. R-DAT Block Format

	CXA 1045Q	CXA 1046M	CXD 1008Q	CXD 1009Q	CXD 1052Q
STRUCTURE	BI-POLAR	BI-POLAR	CMOS	CMOS	CMOS
CHIP SIZE (mm)	3.23 X 3.9	2.64 X 2.64	6.79 X 7.01	8.73 X 8.03	7.11 X 7.04
POWER SUPPLY	+5V	+5V	+5V	+5V	+5V
POWER CONSUMPTION	90mW	80mW	70mW	100mW	30mW
PACKAGE	48-pin QIP	28-pin MFP	64-pin QIP	80-pin QIP	48-pin QIP
FUNCTIONS	• RF REC / PB • PB EQ	• ATF PROCESSING	• ERROR CORRECTION • A/D/A INTERFACE	• MODULATION, DEMODULATION • MEMORY CONTROL • SUB CODE PROCESSING	• DRUM SERVO

Table 5. R-DAT — Sony Chip Set

- ERROR RATE OF PRINTED TAPE
  THE AVERAGE BLOCK-ERROR RATE BEFORE CORRECTION IS LESS THAN 10^{-3} AND THE REPRODUCED SIGNAL IS FULLY CORRECTED AND STABLE.
- ENVELOPE FLUCTUATION OF PRINTED TAPE
  ENVELOPE FLUCTUATION OF SIGNAL OUTPUT WAS 1.5dB P-P, LOWER THAN THAT OF CONVENTIONAL HEAD-RECORDED COPY TAPE.
- TRACK PATTERN LINEARITY OF PRINTED TAPE
  THE TRACK LINEARITY OF PRINTED TAPE WAS LESS THAN 3 μM P-P.
- SPECIFICATIONS OF THE PRINTER
  BASIC SPECIFICATIONS OF THE PRINTER:

TAPE WIDTH	3.81 (MM)
BIAS HEAD	FERRITE RING HEAD
FREQUENCY OF BIAS FIELD	200 (KHz)
CONTACT PRESSURE	4 (KG/CM)
PRINTING SPEED	4 (M/S)
REWINDING SPEED OF MOTHER TAPE	6 (M/S)

Table 6. R-DAT — High Speed Printing — Printed Tape Characteristics

Operational Experiences and Future Expectations from the M-II Format

Steven Bonica
National Broadcasting Co.
New York, New York

The choice of the MII format for use at NBC and the reasons which led up to the choice are discussed. The format and the initial specifications are reviewed. The product line is described, and recent installations for network delay and News operations are detailed. Planned installations for VT Operations and future applications, including automation, are also addressed.

I'd like to share with you some of the early results of NBC's quest to implement a video tape recording system with the broadest possible range of application. Our quest for a new universal solution was based on a strategic initiative. We had determined that if a videotape system could be designed which possessed a certain set of attributes, it could potentially provide our industry with applications extending beyond any system in the past.

FUNDAMENTAL REQUIREMENTS

A simple, but unique list of fundamental requirements is the foundation for our entire program.

1. First, the system must be cassette based to allow for automation and ease of tape handling.

2. The system must have a lightweight fully featured camera recorder.

3. The system must exhibit picture quality, and attributes comparable to C format.

4. Finally, the system must have at least 60 minutes capacity to allow for automation of program length material.

SMALL TAPE PHILOSOPHY

What we then tried to do, was to develop a philosophy for utilization of small tape. We felt there was potentially a wider range of applications for a 1/2" cassette based system than current thinking indicated. Possibly the system could provide for all the needs of television production, except perhaps high end postproduction and graphics.

A complete family of products was desired. From portable equipment to studio equipment and multi-cassette machines.

The system had to maintain the features that were currently used for each of the individual applications. For example, we required 60 minute capability, for the ENG application. Additional audio capabilities were also desired to make possible stereo field production, and to add additional flexibility to other production operations.

A cassette based system clearly brings with it the capability for automation applications. Multi-cassette machines with program length capability can be applied to a wider range of broadcast operations than has been previously possible.

Mobility for news gathering operations has always been a problem. There was an increasing need for more compact and lighter systems for doing extensive editing in the field.

SMALL TAPE TASK FORCE

The next step was to establish a set of requirements which detailed our needs. Our technique was to form a Task Force to determine the necessary requirements for a replacement format. This group consisted of representatives from News, Owned and Operated Stations, Sports, and Production. The Task Force produced a document which was our reference for evaluation of system capability. This included detailed specifications, and product availability.

EXAMINING THE OPTIONS

When we examined the immediately available offerings, we found that they did not provide what we desired. In an attempt to research the matter further, discussions with broadcasters and many manufacturers were held to ascertain their ideas about future developments. What we learned from NHK in Japan was that they had similar goals to ours, and had begun development of a 1/2 inch tape system to replace quad machines. Our analysis of the situation indicated to us that our goals were attainable.

AUDIO & VIDEO SPECS

The specifications and equipment features were then submitted to several manufacturers. The specifications shown are basic performance figures and are the same as in the original quotations. Our analysis of the responses, which included technical performance, cost, availability, and other criteria, led us to the conclusion that the MII system was precisely what we wanted.

FORMAT DETAILS

Fundamentally, metal particle tape permits higher recording densities than conventional ferrix oxide tape, and because the tape has a higher output, the carrier to noise ratio is improved. One of the key advantages of this system is efficient usage of metal particle tape. MII uses 43% less than other 1/2 inch metal particle formats, and 66% less than the proposed composite digital format. Obviously this has a major favorable impact on operating costs.

The MII format provides for one luminance track and one chrominance track. The chrominance track consists of two serial time compressed color difference signals. There are four audio channels, two linear and two FM. The FM channels are carried on subcarriers on the chrominance track. A unique time code system is built into all the MII equipment, which provides, among other things, vertical interval time code. Automatic tracking, (Slow-Motion effects) are essential to our operation. This component analog format produces an impressive full-bandwidth picture, at speeds from normal reverse to twice normal speed forward. The cassette format, and some considerable forethought on the part of the manufacturer, makes the system particularly suitable for automation applications.

PRODUCT LINE

The equipment NBC will acquire consists of four key devices.

The camera/recorder is a lightweight robust unit which makes use of a custom designed cassette. We required that the recorder be capable of black & white field playback, as well as confidence playback during record. It was to have the same time code and four channel audio capabilities as the rest of the equipment.

The field recorder/player unit intended for over the shoulder use was specified to accommodate the 90 minute cassette, along with video confidence heads and field color playback. The unit also includes an RF output and full time code capabilities with a character generator. A companion TBC was also requested to provide full broadcast quality video playback in the field.

The studio recorder contains all of the features found in typical studio VTR's, including monitoring and remote control options.

We also specified a field edit machine, which should be a compact, lightweight machine capable of all the studio machine features, except that, the TBC will be an external option.

MOUNTAIN TIME ZONE DELAY SYSTEM

The applications for our small tape systems are turning out to be as diverse as we originally envisioned. Indeed, the Mountain Time Zone delay system was the first installation at NBC, of MII recorders. It was built by National Teleconsultants, and it consists of a rack of AU-600 MII studio recorders and a control computer. It employs no robotic systems as tape exchange is not required for this application. Input and output switching and monitoring are in the adjacent rack. The system was designed to be capable of providing a continuous program delay, which can be varied from 30 minutes up to three hours. Two of these packages were purchased and installed in our New York facility. The MII recorders pictured here are the same recorders supplied to NHK in Japan. These machines pre-date the first AU-650 rolling off the assembly line.

Two redundant systems are used to provide appropriate time delay for programs originating from New York which are then sent to NBC affiliates in the Mountain Time Zone. The recorders are controlled by the computer, which is executing the day's schedule, that was created on an off-line system.

The computer program selects source video and audio, assigns VTR's to record, and at the appropriate delayed time, selects the correct VTR to play back. The computer program assigns machines in a manner so that they all get approximately equal use. It also continuously monitors the operation of the recorders, and, according to predetermined rules, will warn of problems, or will take action to correct them. If a VTR should fail for any reason, it will be bypassed, and the next available recorder will be assigned. These incidents and any other warnings are recorded on a printout for maintenance use. The indications of malfunctions affecting air are remoted to NBC Switching Central where corrective action such as switching to the protection system can be taken.

A monitor in the computer rack displays the status of the system at all times, and indicates the events programmed for the day.

This facility replaces four one-inch type C recorders, which would be required for this task, and which would have occupied at least three times the space, consumed considerably more power, and required a great deal more operator attention.

The system was placed on-line at the end of October.

NEWS DEPARTMENT

In an operation of this size, with source material coming from News, Sports, and various production organizations, it is not possible to effect an instantaneous changeover. Therefore, the video tape facilities will have to accommodate differing formats throughout our transition period. Network News is the next area to introduce MII to its operations. Beginning with New York and Washington their domestic migration will continue into late 1987.

NEWS EDIT ROOMS

Four News Edit rooms in New York have just been completed. They are, Network Edit Rooms EJ2 & 3, and local edit rooms EJ14 & 15. Each room has three MII VTR's and one 3/4 inch U-Matic VTR. Each facility consists of the 4 VTR's, an Edit Controller, a switcher, and an Ampex ADO. The configuration allows considerable flexibility and many special effects.

All four audio channels of the MII VTR's are normalled as sources to a 16 channel mixer. The mixer output will feed the longitudinal tracks of the record VTR. The NBC design includes a cutover switch, so that any two recorders may be operated as an independent edit pair in the event of editor malfunction.

CONVERSION OF BUREAUS

In our simpler straight cuts-only facilities in New York and at the other Bureau locations, Network ENG is converting all of its 3/4 inch U-Matic and hybrid facilities to the MII format. That represents 30 rooms, just in the Manhattan facility alone. In these rooms, the component outputs of the MII recorders will be interconnected, thus the conversion is relatively simple.

NEWS AUTOMATION SYSTEM

NBC will employ several types of automation systems throughout its operations, one of which will provide record and playback of news material in New York. The system will have the capability of handling over 100 cassettes, and it will be possible to add new cassettes at the last minute. This will make it possible to rearrange show routines while on air.

Two of these systems, one for air and one for protection will each include 6 AU-650 recorders. The systems will be interfaced to the house automation and newsroom computers, which will permit pre-programmed record or playback sequences when required. These systems are expected to be on-line within a few months.

VIDEO TAPE OPERATIONS

The M-II format will also be employed in our production operations.

Edit 2

One of our planned facilities is called Edit 2. It will be used by Network Advertising and Promotion. This editing facility is intended to be an on-line or off-line analog component edit room to be manned by a single operator. It will consist of four MII VTR's, a three effects buss component switcher, and a edit controller. It will employ a VCA-equipped, Editor Controlled audio console. Its ergonomic design allows a single operator to manipulate all editing functions, including the loading of VTR's without leaving the control position.

Edit 2 will be connected to NBC's "Video Imaging Systems area" to allow access to all current and future video graphics and special effects devices. Edit 2 is expected to be completed in the first quarter of 1988.

Edit 3

Edit Room 3 is an existing edit facility, which is principally used by NBC Sports. A multi-level video/audio routing system will be added to this room, which will handle NTSC, analog component, video keying signals, and four channels of audio. The room initially will house a complement of five MII recorders, two 1-inch Type C recorders, and two 3/4 inch recorders. Full component operation will be implemented when a suitable production switcher is available.

The audio mixer will be expanded to handle the increased number of audio channels of MII. Special attention will be paid to providing the audio operator with sufficient acoustic isolation and space for high quality critical stereo monitoring.

Multi-Format Dub Room

Unfortunately, our industry will always have to deal with an abundance of video recording technologies, and as the name implies, this multi-format dub room will include a variety of tape equipment to permit dubbing between any formats. The equipment complement will be, ten MII recorders, ten 3/4-inch recorders, four 1-inch recorders, and two 2-inch VTR's. A small complement of 1/2-inch VHS recorders and color correction equipment will be included.

The tape machines will be arranged in clusters, each to be provided with a monitor station. Machine control and signal routing will be centralized. The operator will load tapes and then control multiple machine start, stop, and signal distribution from a single work station.

FUTURE PRODUCTS

Field Edit Machine

With regard to our future planning we must venture back into the field. As part of the small tape program, NBC specified and required the development of a portable field-edit VTR. This machine fills an important gap in providing lightweight full-function recorders for field editing and satellite news gathering operations. It will be available in 1987 and will be utilized by most operating divisions.

Office Viewer

NBC's hardware request also included items such as a low cost player for the office, to allow the production staff greater access to material throughout the production process.

Automation Systems

From the very beginning of this program, a key goal of the system was to provide for the automation of program length material. Our Mountain Time Zone system is the first example of such an application.

Future automation systems will be used to provide playback of all commercial, promotional and program material to the network or local air from New York, Burbank, and the owned and operated stations. They will have the capacity to handle 15 second messages on a continuous basis, and will handle a library of 100,000 messages. They will normally be programmed many hours ahead, but will have the capability of accepting late changes, and accomodating the addition of new cassettes at the last minute.

The largest of these systems will be used in New York and Burbank and will have additional capabilities such as time delay operations, recording from external sources, and dubbing facilities. The system will have the capability of handling up to 10 output playback channels. It will have a library function to keep track of over 300,000 messages. Our plan is that this system will eventually operate in such a manner that a tape can be put into the system, and never be handled manually again until it is removed for purging at some future time.

CONCLUSION

In the past, the question has been asked, how many JND's does it take to make one ERE. That is, how many Just Noticeable Differences equal one Enhanced Revenue Experience.

This program with its sizeable hardware savings, greatly reduced tape stock cost, dramatic manpower and space reductions is creating an E.R.E. of monumental proportions.

In conclusion, I would like to emphasize that NBC's small tape program was a direct result of a significant team effort. Major contributions were made by NBC Sports, Stations, News, Purchasing, and Engineering Divisions. In the team's planning, we laid out some arduous goals. At this juncture, I would like to report to you, there is no doubt in my mind that we will make them.

References.

MII Format VTR — Kunio Sekimoto, Masahiko Matsui, Iwao Obata; IERE,
Sixth International Conference on Video, Audio and Data Recording,
Brighton, England, 18th-21st March, 1986.

Fig. 1. Mountain Time Zone Delay System.

Fig. 2. News Edit Room EJ2.

Fig. 3. News Edit Room EJ2 — M-II Racks.

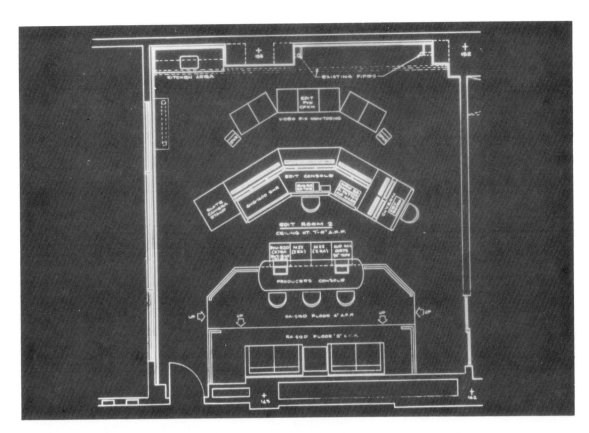

Fig. 4. Edit Room 2.

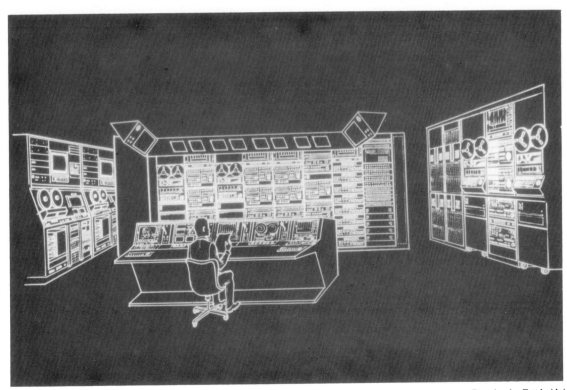

Fig. 5. Multi-Format Dub Room — Artists' Impression.

Steven Bonica is vice-president of engineering at the National Broadcasting Co. (NBC). Prior to assuming his current position, he was director, broadcast systems engineering. Prior to that, he served as manager and then director of videotape operations, network news. Mr. Bonica has been with NBC for his entire professional career. In January 1969, he began in the News Film Dept. as an editing room assistant and later became an assistant film editor, then film editor. During his seven years in news film, his editing assignments took him to Europe, South America, Russia, Africa, and Southeast Asia.

In 1976, he became a videotape engineer in the Engineering Dept. Later that year, he moved to the Electronic Journalism Dept., as videotape editor. Mr. Bonica was a supervisor when he left E.J. in 1980 to rejoin the News Dept. as a technical planning manager. One of his responsibilities in that position was the organization and supervision of videotape operations for both the Republican and Democratic National Conventions.

CBS Experience with Small Format Videotape and the Implications for the Future

Bernard L. Dickens
CBS Engineering and Development Department
New York, New York

ABSTRACT

From the introduction of videotape in 1956, through the beginning of this decade, the broadcast industry has seen relative order in videotape technology. In 1958, two inch quadruplex recording characteristics were quickly standardized to allow tape interchange. Later, the U-Matic tape format became a de facto standard for newsgathering because it was the only small cassette format available. Similarly, the one inch type C system rapidly became a de facto standard by providing important new features, improved performance and lower costs to broadcasters. Today, however, videotape technology sees increasing disorder -- bordering on chaos.

As the broadcast industry has grown, and broadcast technology has become more sophisticated, requirements for videotape formats have diversified. Production, post production, field production and newsgathering all demand unique characteristics and features of a recording system. As these demands increase, improved signal quality and lower costs remain of paramount importance. In an effort to meet the needs of broadcasters, manufacturers have introduced numerous new and incompatible videotape formats during the past five years. Most recently, improved versions of existing formats have been introduced with varying degrees of compatibility with their predecessors. As a result, during the 1980's, broadcasters have been forced to evaluate and choose from among ten different recording systems.

In 1985 CBS introduced a one half inch videotape system in its new Hard News Center. More recently, CBS has evaluated combination camera/recorders for network newsgathering. In this paper the author will report on the CBS experience with small format videotape systems and examine the implications of this experience for the ability of present videotape technology to meet current and anticipated broadcast requirements.

For some time now, a popular topic at conferences, in the trade press and when television engineers meet, has been the proliferation of television tape recording formats and more recently the "Universal Tape Format". The author has participated in many of these discussions invariably advocating minimizing the number of recording formats. The purpose of this paper is to outline the CBS experience with small format recorders. It is first desirable to review broadcaster's requirements as they relate to tape recording formats.

A tape format standard describes the mechanical, magnetic and electrical parameters that are required to ensure acceptable interchange of recorded information. The tape format specification should not be confused with a recorder specification. For example, a low cost recorder/reproducer with very simple manual controls can generate tapes that can be interchanged with those generated by the "top-of-the-line" editing recorder/reproducers. Some of the items that determine the tape format are:

- o   Signals to be recorded
- o   Performance level
- o   Physical parameters of the tape and tape carrier
- o   Need for special features such as slow motion playback.

SIGNALS TO BE RECORDED

The need to record the television video signal whether it is in composite NTSC or component form is obvious. It is also required to record two program audio signals and the SMPTE Time and Control Code. For some applications there is a need for additional audio channels. Some of the more recent formats contain provision for four high quality audio channels; for example, to provide stereo audio and a second language.

PERFORMANCE

The required performance determines many parameters of the format. For analog recording the achievable signal bandwidth is determined by the shortest wavelength that can be recorded on the magnetic tape selected and the speed at which the magnetic head scans the tape. The resulting signal-to-noise ratio is determined by the tape magnetic material, the surface of the tape, the track width and the shortest recorded wavelength. For digital recording the signal bandwidth and signal-to-noise ratio are determined by the maximum bit rate that can be recorded. The bit rate in turn is determined by the same parameters that affect the analog signal bandwidth: wavelength, track width, head-to-tape speed, etc. It can readily be seen that the required video and audio performance has a major impact on the tape format parameters and the recorder's mechanical parameters.

Now let us examine the performance requirements for recording systems. The broad range of applications of television recording can be divided into two fundamental performance categories; those that require head room for further signal processing and multiple generations and those that do not require significant head room. In this context headroom refers to a performance margin beyond that required to reproduce on playback pictures subjectively equivalent to those recorded (grade 4.5 on the CCIR 5 grade impairment scale).

The requirement for headroom is necessary for applications where multiple generations of recording and playback are necessary, as in almost all entertainment programming where signal processing is an integral part of post-production.

For production and post-production the recorded signals are combined and manipulated with ever increasing complexity. We are all familiar with special effects generators that squeeze, zoom, rotate and transform the pictures, as well as chroma keys and matte systems. More recently, layering or compositing have become practical and popular in generating the complex graphics and effects we see today. To accomplish these effects requires manipulating the television signal in component form. Recording systems for this application must be capable of reproducing faithfully the input television signal in its component form without degrading the signal after multiple record/playback cycles.

For newsgathering applications, it used to be said that where the story content is paramount, an intelligible picture would suffice. Today this only applies to the field portion of the news gathering system. Once a news story is received at the studio, it may undergo four or five recording generations prior to broadcast, as increasingly sophisticated graphics and animation are integrated with the news story. Such complex processing requires ample headroom in the recording format.

In broadcast operations where capital budgets limit the sophistication of post-production equipment, a degraded output signal may be tolerated though deplored. In fact, this is simply a challenge to the manufacturer to so improve his productivity that equipment prices can be reduced.

A large number of recorders are being used in applications that require only modest performance head room. Examples of these applications are:
  o   Program delay
  o   Program assembly
  o   Program distribution
  o   Studio output feeds to transmitter

The performance for these applications should be comparable to that currently achievable with Type C recorders. Automated operation with multi-cassette libraries is rapidly becoming mandatory.

We may state then that the high and modest performance requirements can be met with but two tape formats. Clearly, we are far from such a goal.

In a discussion of requirements, the need for very high reliability and very low down time as a result of failures must not be overlooked. These requirements must be met at a competitive cost.

PHYSICAL PARAMETERS OF TAPE AND TAPE CARRIER

The requirement that a magnetic tape be housed in a cassette providing protection when exposed to typical operating environmental conditions and handling has been universally accepted. In addition, use of a cassette is mandatory for automated systems.

The width of the tape and other mechanical parameters of the tape have a direct impact on the size of the cassette for a given play time and an indirect impact on other parameters such as search and shuttle speeds.

NEED FOR SPECIAL FEATURES

Although every recorder designed to meet a particular format standard does not have to incorporate all of the special features that may be required, the format standard does have to include provision for these features. For example, tracking heads are limited in the range of movement and the rate of movement, therefore, if broadcastable slow motion is a requirement, the positioning of the tracks on the tape has to take into account the limitations of tracking heads.

REQUIREMENTS

A simple solution is to design the format for the most stringent application. This approach will unduly penalize the applications with lower performance requirements, especially in the areas of cost and equipment size. As we know from recent history, these penalties are not acceptable to the user. In the early days of magnetic recording, portable TV recorders were large, heavy and required large battery packs for portable operation. In addition, they were expensive. As a result, they were used only for special news events such as the national political conventions. For the bulk of the news coverage 16 and 8mm film was used. When the 3/4 inch U-Matic format was introduced for industrial and educational TV, broadcasters began to use it for news gathering. Electronic News Gathering (ENG) soon became an accepted industry feature. Gradually the U-Matic format penetrated further into the broadcast news operation replacing the more expensive two-inch quadruplex recorders. The penetration of the U-Matic format happened even though it was acknowledged that the video performance of the U-Matic recorders was only marginally acceptable. The U-Matic format was adopted because its small size and low cost filled an industry need.

Today, however, small format recorders of excellent quality are available, but regrettably, there are too many of them.

Does this mean that the industry will always have multiple non-interchangeable formats? For the foreseeable future I believe it does. There is no question that we are moving towards an all-digital world. During the period of change from analog composite to digital components, there will be a need for some transitional component formats. Our goal, however, is to minimize the number of such interim formats, and to secure agreement on a single format for each market segment.

Certainly, there should be one format for sophisticated production and post-production, and a second for most other applications. A single standard for all applications is unrealistic, but the present situation is intolerable. For example, at CBS, we must conduct operations with 7 different formats served by 170 videotape recorders!

WHERE WE STAND TODAY

The digital studio standards embodied in CCIR Rec. 601 is in fact a family of standards which can well meet the two requirements I have outlined, but its introduction will necessarily be slow because of the present high cost of digital recorders and the slow rate at which broadcast plants will be converted to all digital systems.

The SMPTE D-1 format approved last May by the CCIR can record the digital video component signal with no degradation. The D-1 format was covered in depth at the 1986 TV conference and in the SMPTE Journal. The D-1 format can meet the most stringent production and post-production recording requirements. It was hoped that D-1 format recording if used in all recording applications could, because of the resulting large volume, be produced at a cost low enough to be acceptable for most applications;

however, for many reasons this optimistic scenario is not happening, at least not presently.

It is expected that the D-1 Format recorders will replace Type C in the high performance market and will fill new applications for recording which were not previously practical because of recorder performance limitations.

In the remaining market, where Type C is dominant and there are still a few of the 2-inch quadruplex recorders, the situation is not clear. The 1/2 inch Betacam (SMPTE type L), which started as an ENG recorder, has found some application in this market. The improved small format systems, Beta SP and MII, recently introduced, have promise for this market and could also be used in ENG.

The performance level promised by the manufacturers for the improved 1/2 inch formats provides performance headroom similar to that attainable with Type C format recorders. The small format recorders have also been upgraded to extend their playtime to 90 minutes. These recorders appear to be moving into the lower performance market, the market that does not require video performance with large amounts of head room.

CBS AND THE SMALL FORMAT RECORDERS

After a long and careful study, small format recorders are being integrated into CBS's operation in several areas. The small format recorders were chosen for these applications because of their small size, and their performance. Above all they are cost effective for the application. At the CBS Owned stations, Betacam has been rapidly replacing 3/4 inch U-Matic as the ENG field recorder because of its small size, small cassette and superior performance. In most of our Owned stations, the field recordings are returned to the station for editing and then fed directly to air from the edited master. In general the air copy is only second generation. In one of our stations, a multicassette Betacart is used to automatically feed the news stories to air. The Betacam system is finding acceptance at the stations because it meets the stations needs for lightweight, reliable recorders with better performance than the recorders it replaces.

The CBS News Division has approached the introduction of small formats differently. Their various facilities located throughout the CBS Broadcast Center were being completely replaced with a separate dedicated unit to be called the Hard News Center at the time the small formats were introduced. After evaluation of the requirements for the new Hard News Center and the performance of the small format recorders, it was decided to incorporate the Betacam system. The features of the Beta system that were attractive are:

   o   Better performance than 3/4 inch U-Matic
   o   Small size recorder for combination camera/recorders for field use
   o   Component recording prepares for future component studios
   o   Smaller cassette compared to U-Matic and of a later design
   o   Multi cassette system available
   o   2 channel audio plus time code
   o   Lower capital and operating cost

The Hard News Center is a completely new facility in which the recording of remote news feeds, editing, program building, live presentation, and playback to air, are all consolidated on two floors of a new building. This arrangement has led to improved efficiency and flexibility in the preparation of news broadcasts.

The Hard News Center contains the following facilities:

- o   Twenty edit rooms for assembling news stories using Betacam recorders with some 3/4 inch U-Matic capability for file tapes.
- o   Three narration booths for the narration of news stories.
- o   Five edit suites for editing late news feeds, which can then be played back via the control room directly to air.
- o   Two full control rooms equipped with the most advanced video switchers available; stereo, computer controlled audio consoles; and full digital effects. One of these control rooms is used for building multi-segment stories, and integrating them with computer graphics and electronic stills.
- o   An Air Playback/Record Room containing a routing system for 25 simultaneous incoming feeds; Betacart multicassette machines for the automated playback of stories, and Betacam record/playback machines for the second network feed of the news broadcast.
- o   One main studio/newsroom which consolidates all news writers, producers and editors, traffic control and wire services in a single facility. The studio provides great flexibility in on-air appearance, and is equipped with state-of-the-art cameras and a fully computerized lighting system.
- o   One "Flash" studio for news events which mandate the immediate interruption of normal programming.
- o   An International Communications Room which provides for coordination of seven simultaneous incoming international news feeds.

Network news is no longer a simple ENG system -- it is a sophisticated production system.

SUMMARY

To summarize, the view at CBS is that no single tape format can cost-effectively meet the wide range of requirements that exist in the broadcast television industry. It is to the advantage of the users -- the broadcasters -- to avoid a multiplicity of formats that are directed to the same market segment. We should encourage the manufacturers to cooperate on format standards and to compete on technical excellence in implementing the standard format rather than to compete based on the uniqueness of their format.

Manufacturers have much to gain from the economies of scale resulting from the competition in a single large market with common format requirements, while users will enjoy the resulting lower price for equipment.

CBS supports the use of digital components and the D-1 tape format for high performance applications. We look forward to a lower performance digital component format for the more routine applications that do not require large performance overhead. Until such a format becomes available we see a need to replace current Type C format recorders with a smaller cassette system. The improved 1/2 inch systems appear to fill that need.

Bernard L. Dickens is currently employed by CBS Operations and Engineering as senior staff scientist television technology, Engineering and Development Department. Prior to this assignment, he spent eight years at the CBS Technology Center, where he was responsible for studies and developments related to television recording. Before joining CBS, Mr. Dickens was employed by RCA Corp. in a variety of engineering positions in the design and development of magnetic recording equipment for consumer, industrial and military affiliates. Mr. Dickens is a Fellow of the SMPTE and currently holds the position of chairman of the SMPTE Committee on Video Recording and Reproduction Technology. He is chairman of the IEC Sub-committee on Broadcast Video Recording, and is the U.S. delegate for the CCIR Joint Interim Working Party on Digital Television Recording. Mr. Dickens has a BSEE degree from Rensselaer Polytechnical Institute and a MSEE degree from the University of Pennsylvania.

High Performance Half-Inch Metal Tape for M-II Videocassettes

Masaaki Fujiyama, Shigeo Komine, and Koshu Kurokawa
Fuji Photo Film Co., Ltd.
Odawara, Kanagawa-ken, Japan

ABSTRACT

The world's first metal tape for the M-II system was introduced by Fuji
Photo Film in 1985. The main concepts for the breakthroughs were, 1) the
use of a polymer binder possessing high affinity with the pigment to
prevent its aerial oxidation which would lead to demagnetization, and
2) the optimization of the molecular structure of the binder to minimize
the transfer of organic matters to the head, and to control the solubility
of the lubricants in the binder. The binder polymer also enabled the
uniform dispersion of the fine pigment grains to give highly smooth surface
of the magnetic layer.

Ever since the invention of magnetic recording by Poulsen, it has continually, and logarithmically, increased its linear recording density (Fig. 1).   In particular, recent progress in the density is largely due to the use of fine metal pigments.[2]

Meanwhile, the increase of the necessity and urgency of news gathering has been accelerating the use of the VTR systems which are smaller in size and weight.   The examples are 3/4 inch U-matic, half inch Beta-CAM, and also the recently announced half inch M-II and Beta-CAM SP.

Under these circumstances Fuji Film announced the world's first metal tape for M-II system at the 1986 NAB. Here, the authors wish to report the concept for the break-throughs in the development of this tape.

The following items were considered on design of the metal tape :
1.   Susceptibility to aerial oxidation giving rise to serious demagnetization
2.   Susceptibility to output reduction, ultimately leading to head clogging, due to poor abrasiveness
3.   Possible high friction coefficient as a consequence of the smoother coating surface necessary for high S/N ratio
4.   Difficulty in the dispersion of the pigment of a smaller size required to reduce noise

The first problem, demagnetization, was solved by use of a polymer binder possessing high affinity with the pigment.   Metal pigment grains are covered with a thin layer of its oxide, whose formation process and thickness largely influence the rate and extent of demagnetization.   This suggests that a dense oxide layer makes a barrier against attack of air oxygen.   Fig. 2 shows the relatioship between the demagnetization and the oxide layer thickness as determined by Auger electron spectroscopy.

Similarly, the binder molecules adsorb on the pigment grains supposedly to make a barrier against oxygen, as shown by comparison of the demagnetization of the metal powder by itself with that of the pigment dispersed in a binder (Fig 3).   The authors have found that the denseness of the adsorbed polymer on the grains is a key factor of demagnetization and that a certain population of the adsorbing polar group in the polymer chain is necessary to attain dense adsorption.   Fig. 4 shows an example of the relatioship between the adsorption and the population of the adsorbing group.   The demagnetization of the tape thus obtained is small enough for practical use, as shown in Fig. 3.

The second problem, output reduction which ultimately leads to head clogging, is considered to be due to spacing loss caused by gradual accumulation of the organic materials transferred from the tape to the head.   Metal oxide tape is abrasive enough to avoid output reduction while on the other hand the abrasives are not so effective in the metal tape (Fig. 5).   For this reason the metal tape requires such physical properties as to minimize transfer of the organic materials.   The authors intended the optimization of the molecular structure of the binder by regulating the kind and population of hard segments in the polymer chain. In Fig. 6 is shown an example of the mechanical property of the magnetic coating layer.

In spite of very smooth surface, a low friction coefficient was attained by localizing the lubricants to the coating surface. Fig. 7 gives the depth profile of the elements by means of micro Auger-electron spectroscopic data. The ordinate represents the relative amount of the elements while the abscissa expresses the sputter time which stands for the depth. The lubricants contain carbon (C) at the highest concentration, and chlorine (Cl) comes solely from the binder (polyvinyl acetate chloride). Thus it clearly shows that the magnetic layer is not uniform, but the lubricants exist at high concentration on the surface of the coating. (Fig. 8)
The result is assumed to be due to the bleeding of the lubricant as a consequence of the structure formation of the binder polymer triggered by the polar adsorbing groups in the molecule. The envelopes before and after running are given in Fig. 9, showing no practical change even after 100 passes.

Finally, the smoothness of the tape surface is shown in Fig. 10 on behalf of the dispersibility of the fine pigment in the binder of high affinity with the grains.

The authors are convinced that this new metal tape symbolizes the coming new age of industrial recording medium.

Acknowledgment

Suggestive advices of Mr. M. Kawagishi and Mr. T. Niwa, Matsushita Electric Industrial Co., Ltd. are gratefully acknowledged.

References

1)   For general discussions concerning M-II system, see M. Kawagishi and T. Niwa " Characteristics and perform-ance of metal particle tape in a broadcast video recorder ", SMPTE SG, March 1986.
2)   T. Kitamoto, M. Nakamura and S. Takayama, " Magnetic recording tapes for video recording ", SMPTE SG, March 1986

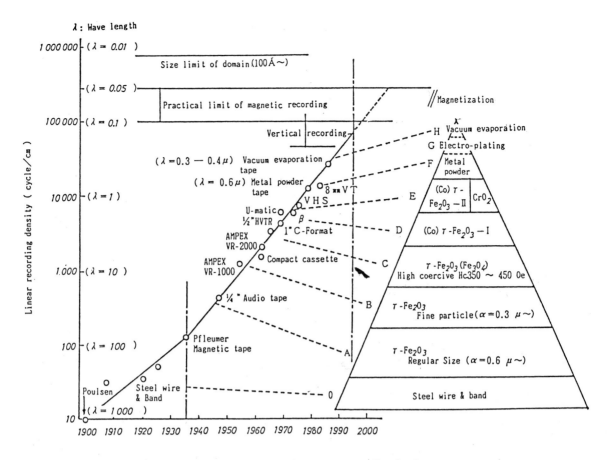

Fig. 1. The progress of the linear recording density.

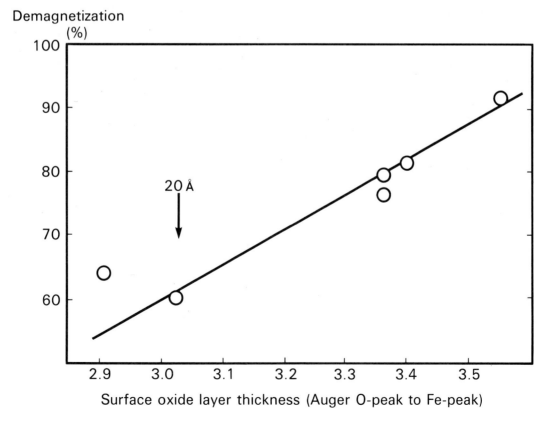

Fig. 2. Demagnetization (60° C, 90% RH, 14 Days) vs. grain surface oxide layer thickness.

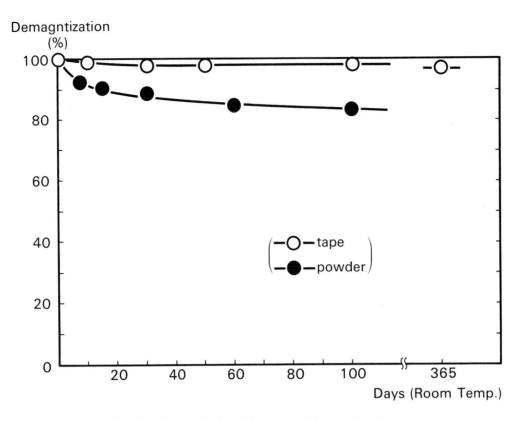

Fig. 3. Demagnetization of the tape and the metal powder.

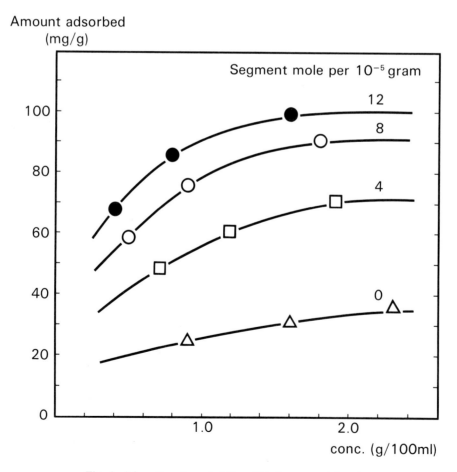

Fig. 4. Adsorption characteristics of binders on metal grains.

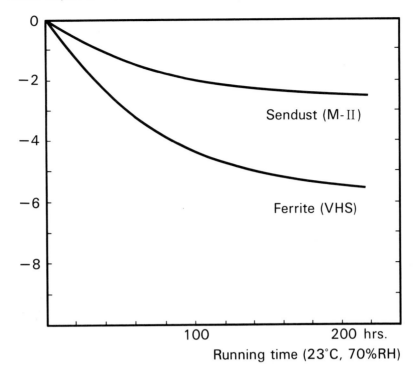

Head depth (μm)

Sendust (M-II)

Ferrite (VHS)

100   200 hrs.

Running time (23°C, 70%RH)

Fig. 5. Difference in head wear of metal and metal oxide tape.

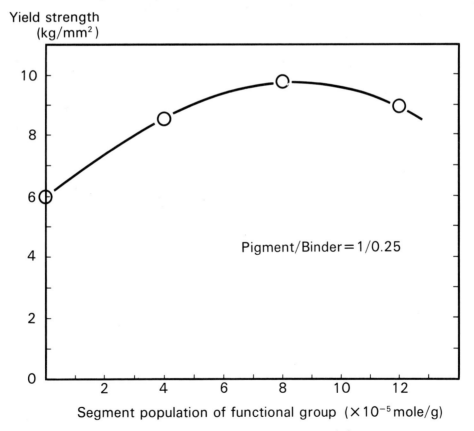

Yield strength
(kg/mm^2)

Pigment/Binder = 1/0.25

Segment population of functional group ($\times 10^{-5}$ mole/g)

Fig. 6. Mechanical characteristics of the magnetic layer.

71

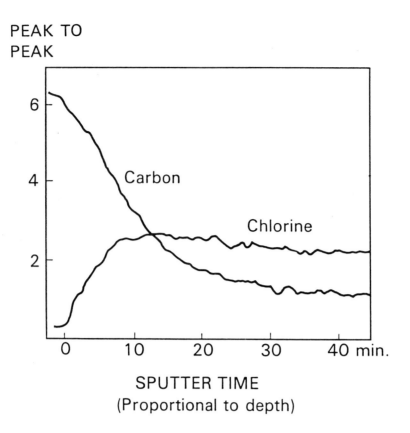

Fig. 7. The depth profile of the elements in the magnetic layer (by means of micro Auger-electron spectroscopy).

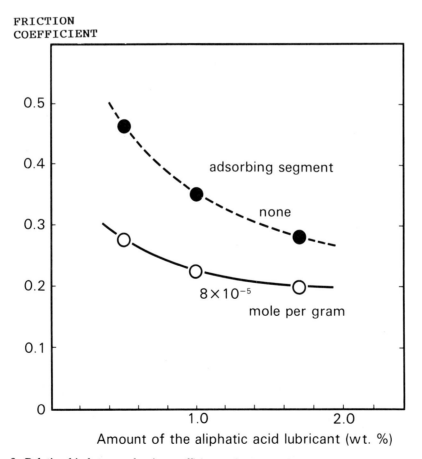

Fig. 8. Relationship between abrasion coefficient and amount of lubricant in the magnetic layer with and without adsorbing groups in the binder.

M II Metal Tape

Magnetic Coating Surface

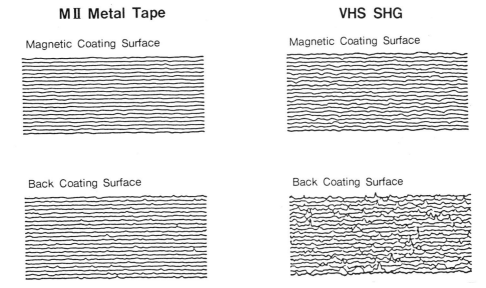

VHS SHG

Magnetic Coating Surface

Back Coating Surface

Back Coating Surface

Fig. 9. RF envelope wave form before and after running.

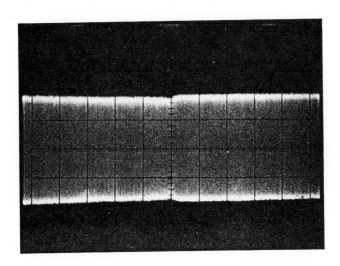

1st Pass

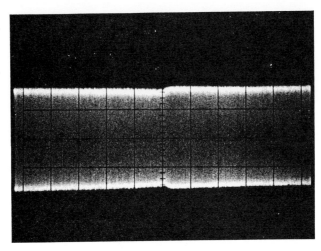

After 100 Passes

Fig. 10. Three-dimensional surface smoothness of the tapes.

Masaaki Fujiyama was born in 1937. After graduation from Numazu Technical High School, Japan, he joined Fuji Film and since then he has been the key member in the development of many outstanding products including SHG, SXG, and 8mm metal tape. In 1984 he was awarded The Chemical Society of Japan Award for Technical Development for the development of the SHG videotape. Present position: Senior Research Staff of Magnetic Research Laboratories, Fuji Photo Film Co., Ltd.

Shigeo Komine, born in 1945, received his master's degree in the field of synthetic and mechanistic polymer chemistry at The University of Gumma, Japan. At Fuji Film, his experience ranges from floppy disk to metal videotape. He was the leader in the development of video floppy disk for the TV-photo system. Present position: Senior Research Staff of Magnetic Recording Research Laboratories, Fuji Photo Film Co., Ltd.

Koshu Kurokawa, born in 1940. After graduation from The University of Electro-Communications, Japan, he joined Fuji Film, and since then he has been working in videotape evaluation technology. Present position: Senior Technical Manager of Magnetic Products Division, Fuji Photo Film U.S.A., Inc.

Development of Component Digital VTRs and the Future Potential of the D-1 Format

Jürgen K. R. Heitmann
Broadcast Television Systems GmbH
Darmstadt, West Germany

ABSTRACT

Since 1979 discussions about digital component recording are under way. For one year now the appropriate equipment has been on the market and further equipment is about to follow. The introduction of the 19 mm component DVTR represents significant steps in terms of technology and operational practice. The first of these is of course the transition from analog video to digital video. Linked to this, however, is a further transition from composite video to component video. The digital component video recorder can be considered as the key for the digital studio of the future.

"Faster than possibly expected, the technique of digital television processing has in recent years entered television studios. This happened smoothly, indeed one might call it hardly noticeable, because digital techniques are still concealed in a black box with analog in- and outputs."

This, ladies and gentlemen, is a quote. It contains the very same words used to introduce a paper which I gave in 1980 on the occasion of the 14th SMPTE Television Conference in Toronto. Seven years have already passed since I spoke at that time about "The Future of the Digital Television Studio".

I think we must concede that what we looked foreward to as the future in 1980 is still seen as the future today. For this reason, please allow me to read a second quote from the paper I gave at that time:

"The answer to the question 'when will a digital studio exist' is less important than the problem of how a current analog studio can be smoothly transformed to the digital future. The large investments involved will allow this to be done on a step by step basis only."

At the very same SMPTE conference, "component coding" was considered by most experts to be the system that would be used in the "ultimate all-digital studio and program production house". The logical result of this was the worldwide digital component interface standard CCIR 601, which was established in 1981. There then followed in 1985/1986 the corresponding VTR standard D-1. What does this standard mean to the user and manufacturer? The main elements of the standard are:

- Digital recording of the video and audio signals.
- Recording of video component signals according to CCIR 601.
- Commonality between 525/60 and 625/50 recorders.
This does not mean that every D-1 component DVTR contains a standard converter. However, the manufacturers are in a position to use the identical tape deck and identical electronics.
- Use of a cassette for the tape.
- Compatibility between recorders made by different manufacturers.

The D-1 standard assures compatibility between recorders made by different manufacturers. In spite of this fact, it still allows the individual manufacturer the greatest possible freedom in terms of mechanical and electrical design. Table 1 shows the development goals for the mechanical part of our D-1 recorder:

1. Robust tape deck mechanism
2. 1 tape deck for all 3 cassette sizes
3. Small scanner having a diameter of 75 mm
4. Replacement of the complete scanner within seconds
5. Control-track head built into the scanner for optimal tracking accuracy.

Figure 1 shows the course of the tape path. The 3 different cassette sizes are indicated in the lower half of Fig. 1. Figure 2 shows the primary target specifications of the BTS component digital VTR. The quality parameters of a digital magnetic recording are no longer given as a signal-to-noise ratio, but rather as an error rate. The value of the so-called "off tape" error rate is of no significance to the user.

The great majority of all errors can be corrected using the built-in error-protection system. The audio and visual quality is determined by the number of errors which cannot be corrected but only concealed. It is our goal to keep the number of error concealments per second to below 200 for the video signal, and below 2 for the audio signal. We expect significantly smaller values, but in the final analysis these would depend upon obtainable tape quality. Naturally, all recorders offer separate head monitoring for video and audio. In conjunction with digital recording technology, this allows the possibility of automatic self-monitoring of the recorder during recording. Digital video recording not only means improved picture quality, but also makes possible extensive automation of the studio, which in the future will lead to significant cost reductions. Although our D-1 DVTR is a highly complex technical device, we also expect cost reductions in terms of maintenance. However, the maintenance procedure is going to be different. The built-in diagnostic system makes fault localization possible up to the printed-circuit board level, and to some degree up to the module level. The diagnostics system may be connected to a personal computer. With the help of special software, it will be possible to refine fault analysis and fault location to a considerable extent. The oscillograph will lose significance in its role as the main means of carrying out maintenance.

We have been discussing digital component recording for 7 years. Appropriate equipment has been available on the market for one year. As I have reported, more is soon to follow. In spite of this, there does not seem to be an end in sight to the discussions, questions and uncertainties among users and manufacturers. New systems of recording, such as "component analog" or "composite digital", contribute to this. And many experts see High Definition Television on the horizon. Which path should users and manufacturers follow? What kind of a future does "component digital" offer?

Introduction of the 19 mm digital tape recorder in the D-1 format signifies several significant steps in regard to technology and "operational practice". The first of these steps is naturally the transition from analog video to digital video. This is nothing new. Digital equipment has long been a familiar sight in television studios. For example:

- Timebase correctors
- Synchronizers
- Standard converters
- Telecine, such as our FDL 60, with digital store
- Digital slide libraries
- Digital video effects
- Digital graphics equipment
- Noise reducers

It should be noted that most of these devices not only function internally in a digital mode, but in a "component digital" mode. The digital component video recorder completes the chain. It can be seen as the key to the completely digital studio of the future.
However, it is not sufficient to solely consider the potential quality of the CCIR 601 interface code or of the digital recording, if in doing so we overlook the significance of the surroundings. Incorrect methods of introducing this new equipment are accompanied by the risk of throwing away all the advantages of processing in the digital component domain. In addition, we must bear in mind that the only sensible introduction of new technology is an evolutionary introduction.

The main step into the future is not the transition from analog to digital. It is the transition from composite to component. New half-inch analog recorders, such as the Betacam-SP, are not the opposite of the digital component recorders. These analog recorders offer a picture quality superior to that of the one-inch B and C formats - that is, as long as processing remains in component technique. At the very least, this requires installation of small editing suites in component technology. When in the course of studio signal processing a component signal is converted into a composite signal and vice versa in order to be able to use existing system components such as the analog composite mixer, the result is a significant loss of video signal quality.

Signals from different video sources are subjected to a multitude of processing steps in the course of a studio complex. In the digital studio of the future - which will be the environment for digital D-1 component recorders - it will be exclusively component signals that are to be processed. The following system components therefore must be available in digital component technology:

- Signals from picture sources
- Vision mixer/switcher
- Interface equipment to/from analog NTSC
- Interface equipment to/from MAC-signals
- Effects equipment
- Synchronizing system
- Routing system
- Monitoring display
- Diagnostic system

Picture source signals, such as from a camera, telecine or graphics generator, generally already produce signals in the component format which are available in parallel form as R, G, B or Y, Pr, Pb.

The vision mixer, which consists of the source selection matrix, the mixing section with effects etc., needs a multi-channel design for the component signal. It is already available in analog form.

In conjunction with analog component recorders, there results the following minimal configuration of an editing suite shown in Fig. 3. At first studio areas using NTSC will exist parallel with areas using component technique. That is why it is necessary to prepare a transition between these two areas. This process can be carried out with great precision using adaptive controlled 3-dimensional comb filters.
In the case of higher demands in terms of video or audio signal quality, the analog component recorders can easily be replaced by digital D-1 video recorders equipped with analog-to-digital and digital-to-analog converters at their in- and output interfaces (Fig. 4).

If we go one step further and replace the analog mixer with a digital mixer, we then have the core of the digital studio (Fig. 5). This is a significant step not only in terms of technology. It is also an important step in regard to daily operational practice. Video signal quality, and above all audio signal quality, are no longer determined by the number of recording generations necessary for a production. This can also lead to use of simpler, and thus more costefficient, video mixers and effects devices.

Thus no longer must all desired signal manipulations be carried out in one step, but can rather be carried out one after the other in the course of a further recording generation, and without quality loss.

The advantage of all devices in composite NTSC technique, and thus of a composite digital recorder as well is the fact that it can be simply integrated into existing facilities. A composite recorder is well-suited in cases in which the task always consists of recording signals which are present in the NTSC form and of reproducing them without further processing. It appears to me to be questionable whether a composite recorder must be digital for this purpose, since the advantages of digital recording first become apparent during the production of programs. A high number of recording generations results here. But many production devices function only or better in component technology, such as the chroma key or digital effects. Every user must decide for himself which path is most advantageous: component or composite.

What significance does HDTV have in regard to this decision? The HDTV production standard will be a component standard. One day, there will be a HDTV broadcast service directly to home viewers. But remember the beginnings of NTSC color TV. It was heavily mixed with black and white material. Assume the similar for an early HDTV service, which may be mixed with standard TV material. The presentation of a CCIR 601 signal converted to HDTV on a wide screen is clearly superior to a converted NTSC signal.

Signal Distribution in the Studio with Component Signals

Use of component signals in the studio makes new methods of signal distribution between the system components necessary. Todays methods of signal-distribution via a single coaxial cable is characterized by a high degree of flexibility and economic efficiency. This is especially true for the switching matrices, the distribution amplifiers and cable equalizing. This technique can be transferred to the component signal format without a great degree of modification.

Currently, possible distribution formats for component signals are:

- Digital component, parallel format CCIR 601
- Digital component, serial format
- Analog component, 3 cable technique (Y, Pr, Pb or R, G, B)
- Analog component, 1 cable: time-division multiplexed (MAC)
- Analog component, 1 cable: frequency-division multiplexing (FDM)

Digital parallel distribution and analog 3-cable distribution are very well suited for signal distribution within a room or within a very limited studio complex. The advantage to this is that the signals are available with no undue difficulty as separate components, which is necessary for numerous applications. An evolutionary development of the present-day analog studio into the digital studio of the future, with its component signal structure, requires the coexistence of digital and analog devices. In general, "component signal processing" makes necessary the parallel availability of signal components in their original time-base. For a TDM or MAC signal, this means reconversion into a parallel format while undergoing simultaneous signal time expansion.

Even in the simplest case of pure TDM signal transmission, every signal component will be subject to two format changes (A/D and D/A) and four filter processes. It is entirely clear that multiple cascading of the conversion process leads to an undesired loss of signal quality. This speaks in favor of an analog component signal distribution format, assisted by frequency division multiplex technology. But let us discuss the bit-serial transmission of the digital component signal. Parallel distribution of the digital component signal is only suited for distances of up to a maximum of 300 m. Greater distances must be linked by means of bit-serial transmission via a 75 ohm coaxial cable with BNC plug connectors. The standardization of this type of signal format seems to be completed. The base is a so-called 8/9 code. The 8-bit parallel data words are first converted into a 9-bit word. It is only then that conversion into a serial data signal is carried out; the data transfer rate of the signal no longer corresponds to that of the CCIR 601, which is 216 Mbit/sec, but rather has increased to 243 Mbit/sec. The 8/9 code thus offers extensive DC freedom. The disadvantage is that in addition to the unnecessary increase in the data rate, the spectral energy maximum of this signal is at $f = 121.5$ MHz - this is the international distress frequency for aviation.

I don't completely understand why a 8/9 code should be necessary for signal transmission via a coaxial cable, when the digital recording in the D-1 format with the significantly more difficult magnetic recording channel is carried out using an 8-bit code. I would prefer that the selection of codes for serial cable transmission be carefully reviewed one more time. Alternative proposals are available and I have heard that demonstrations are to be given as part of a parallel program to this conference.
The advantage of bit-serial transmission is the two conductors per source transmission via a single coaxial cable. However, in order to attain this a serializer, a deserializer and an equalizer are required. Should it be possible to pack these circuits into one IC each, then every digital device could be equipped accordingly. Connecting of digital component devices could be handled as simple as the connecting of analog NTSC equipment. Even connecting analog component devices may then be done via digital bit serial transmission using additional A to D and D to A converters. The transition from analog components to a single-cable digital signal will then be less expensive and of higher quality than the generation of an analog single-cable TDM signal.

Looking Towards the Future

Let us return to the digital D-1 recorder in the component studio.

The full benefits that result from the advantages offered by the digital component recorder can only then be realized when a larger range of new equipment, such as the vision mixer, or signal distribution etc. is available as part of the environment.

It is realistic to assume that during a transition phase the environment of the digital D-1 recorder will not initally operate in the standardized digital signal format. As a result, there will be different devices using analog component technique to be integrated into the digital signal path. Acceptable, temporary solutions for this problem exist.

However, careful planning is necessary in order to ensure that the high level of quality offered by the digital D-1 video recorder because of its component signal processing is not lost as a result of to many compromises in the environment.

Never before existed an SMPTE-VTR-standard prior to the first market introduction. This is the success of the user, represented by their international organizations SMPTE and EBU. The new D-1 recording format takes a manifold of user demands into consideration. All manufacturers started with digital composite recording. We changed the way to component recording due to the important influence of the international standardization work. The acceptance of this D-1 format will ultimately decide whether this kind of standardization procedure with decisive influence of SMPTE and EBU will prove a once in lifetime happening or not.

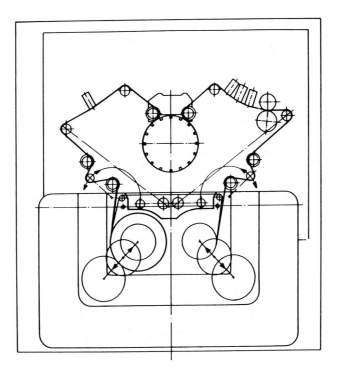

Fig. 1. Tape Transport.

1. Video acc. to CCIR 601

2. 4 independent digital audio channels acc. to AES

3. Number of concealed erroneous pixels < 200/sec

4. Number of concealed erroneous audio samples

   < 2/100 sec

5. Independent editing of video and 4 audio signals

6. Confidence playback for video and audio

7. Built—in self testing procedure

8. Picture at non standard speeds (+/−): 0...0.5, 2...25

9. Pictures at speeds: 0.5...2 (option)

10. Read—modify—write procedure for audio (option)

Fig. 2. Design Objectives for the D-1 Electronics.

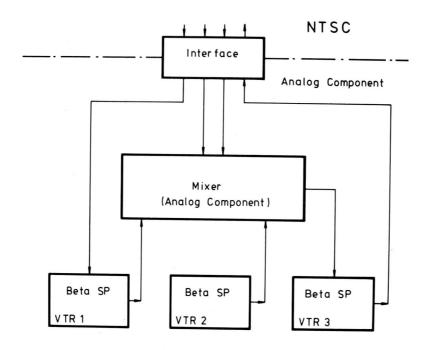

Fig. 3. Analog Component Editing Suite.

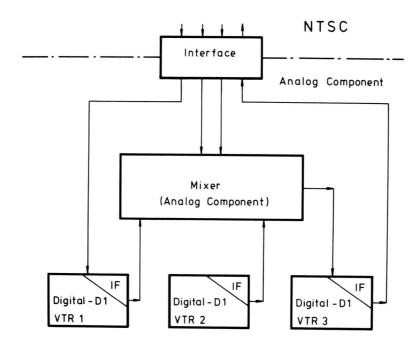

Fig. 4. Hybrid Component Editing Suite.

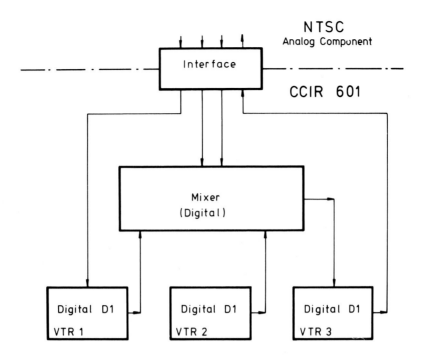

Fig. 5. Digital Component Editing Suite.

Jürgen K. R. Heitmann was born in Hamburg, Germany, in 1946. He received the "Diplom-Engineer" degree in communications engineering from the Technical University of Braunschweig, Germany, in 1972. He joined Robert Bosch GmbH, Television Studio Equipment Div., in 1973, which is now BTS—Broadcast Television Systems GmbH. Mr. Heitmann was responsible for the advanced development of digital video recording. Since 1986 Mr. Heitmann has been department manager for magnetic recording systems and in this position he is responsible for all product and advanced developments in magnetic recording. He is a member of the SMPTE and a participating member the SMPTE Working Group on DTTR and the Committee on Video Recording and Reproduction Technology.

The Composite Digital Format and Its Applications

Edwin Engberg, Richard Brush, Maurice Lemoine,
Steve Magnusson, Fraser Morrison, Dave Rodal,
Dennis Ryan, John Watney
Ampex Corp., Audio Video Systems Div.
Redwood City, California

ABSTRACT

With the introduction of professional digital video tape recorder hardware, much discussion has arisen in the video community over the issues surrounding component and composite digital recording methods. Some believe the two recording methods will battle in the marketplace with one becoming the eventual winner. We at Ampex believe both formats will be in use over the next decade with the specific features and benefits of each format matching different applications.

This paper describes the composite digital format and highlights its major features. To technically understand the format a description is presented which contains information about:

    1.   Encoding parameters
    2.   Mechanical parameters
    3.   Record content
    4.   Longitudinal tracks
    5.   Media

The important differences between the composite digital and D-1 formats are compared. Emphasis is placed upon how the composite digital format meets the needs of the user who records and distributes video in composite form.

SECTION 1 -- INTRODUCTION

With the introduction of professional digital video tape recorder (DVTR) hardware, a discussion has arisen in the video community over the issues surrounding digital component and digital composite digital recording methods. Some believe the two recording methods will battle in the marketplace with one becoming the eventual winner. We at Ampex believe both formats will be in general use over the next decade with the specific features and benefits of each format matching different applications.

This paper describes the composite digital format, compares it to the D-1 component format and explains the benefits it brings to the professional video marketplace. The 525 NTSC version is used as an example for this paper.

Historical Perspective:

Digital video recording has been publically demonstrated for over 10 years. The engineering development over this time has been focussed upon making the technology practical and meeting evolving operational requirements. This has led to the development of both composite and component recording techniques with each providing operational advantages for segments of the marketplace.

The first DVTR was shown in the mid 70's by John Baldwin of the IBA. This was followed in the late 70's and early 80's with demonstrations by Ampex, Bosch, Hitachi, NHK and Sony. Most of this early work was done recording composite television signals.

In the spring of 1982, the CCIR approved recommendation 601 which is a transmission standard for a digital component signal. Both the CCIR and SMPTE have working groups which have established a digital recorder standard (D-1) based upon this component 601 recommendation.

In recognition of the real need for both component and composite digital recording standards to meet different applications within the television industry, at NAB 1986 Ampex introduced the ACR 225 commercial spot player which is based upon the new composite digital format.

There is a desire within the world wide professional video community for a single universal format that meets all recording needs. The economics of this is very attractive. The problem is to put the capabilities for all of the necessary operational and performance features together into one universal format: High quality audio and video performance; Transparent multi-generations; Small size and light weight; Low power consumption capability; Cassette-ability; Low media cost; 2 hour or longer record/play time; Complete editing capability; Graphics compatibility; etc., etc.

DIFFERENT APPLICATIONS -- DIFFERENT REQUIREMENTS

A universal format has been impossible to accomplish in the past, therefore to meet the important requirements of different applications, different formats were created. Examples of professional applications that today require significantly different and conflicting format solutions are: ENG which needs light weight and low power; production and post production which needs high quality video and audio performance with multigeneration capability; the industrial recording segment which needs low cost equipment and media. It still appears that for the foreseeable future a single format that meets all application requirements is an impractical.

THE NEXT STEP - DIGITAL

Formats that will be accepted for general recording applications in the future will provide a step upward in both performance and return on investment for the user. Digital recording offers potential for new and increased operational features, improvement in signal performance and reduction in acquisition and operating costs. These benefits may justify the general acceptance of new format standards. Today there are two format choices available for digital recording, the 4 times subcarrier sampled composite and the D-1 component.

THE PROPER MATCH

There is an important factor to consider when evaluating the optimum digital recorder format for an application. It is the video signal usage and distribution environment. Presently, transcoding between composite and component signals with economically practical transcoding hardware causes noticeable deterioration in video quality. This is true whether the transcoding is done by analog or digital techniques. Therefore it is strongly advisable to stay in one video signal form.

Composite Match:

Currently the broadcast studio operates in an almost completely composite environment. Therefore there is little necessity within such broadcast plants to operate in components. Ease of signal distribution and signal control as well as the relatively high cost of digital component implementation indicates that broadcast distribution will remain composite analog for some time to come. A composite digital recorder is well matched to this environment.

Most editing applications require high quality video and audio through several generations. Here a composite digital recorder can directly replace the analog recorder, without obsoleting the other equipment in the editing system. The digital composite recorder can fit directly, plug for plug, and provides as many as 20 transparent generations.

Component Match:

There are some applications where a component recorder is a proper match. Some post production islands doing manipulation of graphic images that require many generations would benefit if they were interconnected in digital component.

Another possible component application is the transfer of a video program between different transmission standards. Analog and digital component recorders have an advantage here. Transcoding between two standards with the same scanning frequencies is made much easier if the video signal is available in components. Even transcoding between 525 line and 625 line standards is easier with a D-1 digital component recorder because only the field frequency difference (number of scan lines) needs translation.

COST

The final deciding factor will inevitably become economic. What is the return on investment in a format for an application? What is the initial investment cost for the equipment? How much is the re-occurring cost for the recording media and labor to support the equipment?

For one application, integration into a digital component island may be more important than the initial acquisition cost of the equipment. For another, continued utilization of existing analog composite equipment surrounding the VTR and re-occurring costs may be the deciding factors.

For the specific example of composite versus component digital recorders, the cost to manufacture the D-1 component recorder will be higher than that of a 4fsc composite digital recorder. This almost always translates into a higher price to the buyer. The difference in cost is fundamental to the need of the component digital recorder to record and process a significantly higher data rate signal as compared to a composite digital recorder. This make the signal system larger and the scanner more complex.

Also, the resulting lower recording data rate required of the 4fsc composite digital recorder is one factor that helps it to record almost 3 times as much on a cassette as a D-1 component recorder. These differences can directly translate into lower equipment acquisition cost and lower cost per hour of recording time for a composite digital recording system compared to a D-1 recording system.

A CHOICE IS NEEDED

Most all recording applications will benefit from the performance of a digital recorder. Currently, not all applications need nor can afford the support equipment and higher cost of a digital component recorder and system. The choice between digital component and digital composite recording is needed to most efficiently and economically match the digital recorder to the application.

SECTION 3 -- FORMAT DESCRIPTION

The following detailed description of the composite digital format is meant for those who want to know about the inner make-up of the format. Most readers will gladly skip to the next section which compares it to the D-1 format.

The composite digital format provides:

One digital video channel

Four independently editable digital audio channels

One analog audio cue channel

One time code channel

The video and audio signals for the digital channels may be input and output in either analog or digital form.

VIDEO ENCODING PARAMETERS

Table 3.1 summarizes the encoding parameters for the NTSC version.

The analog composite color signal is encoded in composite form. Sampling is at 4 times subcarrier, using 8-bit linear quantization.

Information during the horizontal blanking interval and color burst are not recorded on tape. Appropriate blanking and burst data are recreated for output during playback.

768 samples per line are recorded, centered about the active picture. 255 consecutive lines from each field are recorded in 3 segments of 85 lines each. Each segment (1/3 of a field) is recorded within a pair of adjacent helical tracks. Samples are distributed within the segment (helical track pair) in a pattern which alternates from line to line.

AUDIO ENCODING PARAMETERS

Audio sampling rate is 48.00 kHz and is locked to the horizontal frequency of the video signal. Resolution of each analog audio sample is 16 bits with provision in the format for 20 bits to match the AES recommendation. Coding is two's complement linear PCM.

HELICAL TRACK CONTENT

Helical tracks contain digital data from the video channel and four audio channels. Figure 3.1 and Table 3.2 show and describe the on-tape magnetic footprint as observed from the magnetic coating side.

Audio data is contained in four recorded sectors per helical track, two at the beginning, A_0 and A_1, and two at the end, A_2 and A_3. Audio data is recorded twice, once on each track of the track pair, at opposite ends. This minimizes uncorrectable errors during reproduce.

Video data is recorded in the large sector in the middle of each helical track. Six helical tracks make up one TV field. Two helical tracks (track pair) make up one segment.

An edit gap between each audio or video sector accommodates timing errors that can occur during editing.

Helical Track Arrangement:

Each helical track, as shown in figure 3.2, is divided into the following elements:

1. Four audio sectors and one video sector.

2. A start of track preamble (T) of 62 bytes, preceeding the first audio sector.

3. A preamble (E) of 28 bytes, preceeding each of the remaining sectors.

4. A postamble (P) of 6 bytes, following each sector.

5. An edit gap of 156 bytes nominal, separating adjacent sectors.

Each sector in the helical track is composed of sync blocks; 6 for audio sectors and 204 for video.

Sync Block Form:

Figure 3.3 shows the sync block form which is common to the audio and video sectors. Each sync block contains a 2 byte sync pattern, a 2 byte identification pattern, and two inner code blocks each containing 85 data bytes (outer error correction code check bytes are considered data) plus 8 inner check bytes. In addition, the ID is part of, and protected by, the first inner code block.

Sync Block Identification:

The sync block identification pattern is shown in figure 3.4. The first byte of the sync identification pattern, ID_0 (byte 2 of the sync block), identifies a unique sync block within either the video sector or the audio sectors. The sync block numbers are arranged to follow a coded sequence along the helical track.

The specific helical track in which the sync block is contained is identified by the second byte of the identification pattern, ID_1 (byte 3 of the sync block). The least significant bit of this byte (V/A), identifies the sync block as being in a video or audio sector.

The video sector identification pattern field number follows a modulo 4 sequence, with segment zero of field zero coinciding with a control track color frame pulse.

The audio sector identification pattern segment number coincides with the video sector segment number, however the audio field number follows a modulo 5 sequence. This is due to the fact that in NTSC, the number of samples in each audio block or sector varies over a 5-field sequence. The audio 5-field sequence may start at any video field, but once started, continuity of the sequence is maintained, including through edits.

In addition to the sync and identification patterns contained in the sync blocks, each preamble and postamble on the helical track contains a sync pattern and unique identification pattern, to enhance sync detection and acquisition.

Edit Gap:

Space between sectors on a helical track, exclusive of postamble and preamble, is nominally 156 bytes long and is used to accommodate timing errors during editing. During an edit, the edit gap of an edited sector will naturally be partially rewritten.

ERROR CORRECTION

A concatenated Reed-Solomon product code is used for error correction. This consists of an inner code block form that is the same for both audio and video data and an outer code block form that is different for the audio and video data. The inner and outer code for both audio and video operate in the Galois field GF (256).

Inner Code Block:

The inner code block form, shown in figure 3.3, is common to both audio and video sync blocks. The first of the two inner code blocks, block 0, contains 95 bytes consisting of the two identification pattern bytes (ID_0 and ID_1), 85 data bytes (B84 through B0), plus the 8 inner error correction code (ECC) check bytes. The second inner code block, block 1, contains 93 bytes instead of 95, with 85 data bytes plus 8 inner ECC check bytes.

Each inner code block is capable of correcting up to 3 bytes (3 video samples) in error and detecting more than 3 bytes in error with a high degree of confidence.

Video Outer Code Block And Product Block:

Each video outer code block contains 68 bytes, consisting of 64 video data bytes and 4 outer check bytes. Each outer code block is capable of correcting up to 4 bytes (4 video samples) in error, using "erasure" pointers from the inner code decoder. An erasure is when an error's location is known but its value is not.

Figure 3.5 illustrates the video product block sector array, functionally positioned between the inner and outer coders, necessary for the concatenated Reed-Solomon product code. The product code is capable of correcting a burst error of up to approximately 2000 video samples long, corresponding to a time of about 240 microseconds or about 3 TV lines.

Audio Outer Code Block And Product Block:

Each audio outer code block contains 12 bytes, consisting of 8 audio data and status information bytes plus 4 outer check bytes. Each outer code block is capable of correcting up to 4 bytes in error, using erasure pointers from the inner code decoder.

Figure 3.6 illustrates the audio sector product block array necessary for the audio product code form. The product code is capable of correcting a burst error of up to approximately 340 bytes (85 audio samples), using only a single copy of the audio block. In practice, double recording of the audio block within the segment (track pair), provides a much greater degree of correctability.

Video Sample Shuffling:

The video data in each helical track is shuffled before being written to tape. The shuffling distance is over all the television lines within a segment (85 TV lines). Outer error correction check data is not shuffled, but is recorded at the beginning of its video sector on tape.

The shuffling algorithm may be considered as an intra-TV line shuffle process, followed by the outer ECC coder, followed by a sector array shuffle process.

The nature of the shuffling algorithm is such that the samples from one or more sync blocks on tape will appear more or less uniformly distributed within an 85-line (1/3 field) segment of the TV screen.

Intra-TV Line Shuffle:

Each television line contains 6 outer code blocks. Samples within each outer code block are spaced 12 samples apart within a television line, although they are arranged in a permuted order within the outer block. The horizontal sample number of the first sample in each outer block is given by an algebraic function which depends on the TV line number and outer block number within a TV line. The sample number increment between consecutive samples within an outer block is a constant which generates a permutation of the samples, spaced 12 apart within a television line.

Sector Array Shuffle:

The sector array shuffle is a permutation which results in each inner ECC code block containing one sample from each television line within the corresponding 85-line segment. Shuffling operates identically for all video fields and segments.

The shuffling algorithm for the two helical tracks of a segment is slightly different, which produces an offset between track 0 data and track 1 data, and gives additional protection against scratches or defects which may affect adjacent tracks.

Audio Segments:

Audio data is processed in segments corresponding in duration to one third of a video field. Each segment contains approximately 267 audio samples for an audio channel with associated AES status, user, and validity data. In addition, a number of control and user words are added.

Audio Block Layout:

As shown in figure 3.6, each time segment of audio data is processed into an audio block of dimension 12 x 85 bytes, which corresponds to an audio sector on tape. The data portion of the block is 8 x 85 bytes. Four rows, 2, 4, 7, and 9, contain the outer error protection data associated with each column.

Inner protection code, sync block form and channel code are shared in common with the video data.

SCANNER REQUIREMENTS:

In a manner similar to other recent VTR formats, the specification philosophy for the composite digital format has been to identify and tolerance the on-tape magnetic footprint rather than the specific mechanical design of the VTR. This provides flexibility to each manufacturer in the design of the head/scanner/transport combination used to produce a composite digital format recorder.

One implementation of a scanner for this format is illustrated in Figure 3.7 and specified in Table 3.3. By using four record pole tips in combination with an active wrap angle slightly less than 180°, efficient two data channel operation is achieved.

By grouping the pole tips into pairs, the number of automatic scan tracking (AST) and fixed type support structures required can be minimized.

Noticeably absent from this design are flying erase heads. This greatly simplifies the scanner by eliminating 1/3 of the total pole tips and their associated components.

SECTION 4 -- COMPARISON TO D-1

The best way to explore the similarities and differences between the composite digital and D-1 formats is to focus upon the important functional characteristics and hardware implications of the two formats.

Format Application:

Table 4.1 contains the basic application requirements for both formats. The D-1 format was created by committee to meet the need to record CCIR 601 digital component video signals. Some attention was given to the cost of the resulting equipment but the overriding concerns were for function and performance.

The composite digital format was created to meet the functional and economic needs of the professional video recording community who now and for some considerable time will use and distribute video in analog composite form.

Record/Play Time:

The D-1 family of cassettes was adopted for the composite digital format because of its excellent design and the work already underway by manufacturers to develop D-1 cassette production tooling. Dependence of the D-1 format upon the medium and large size cassettes to get 30 and 90 minute record/play time was seen as a costly penalty for the user. Therefore the composite digital format's basic requirement for record/play time was set to 30 minutes with the small D-1 cassette, 90 minutes with the medium and over 200 minutes with the large.

DATA SAMPLING PARAMETERS

Video Sampling:

Table 4.2 gives the comparison for the data sampling parameters. Higher sampling rate in the composite digital format gives slightly better luminance bandwidth. Separate chrominance sampling in the D-1 format gives a slightly wider chrominance bandwidth and a slightly better chrominance signal to noise ratio.

Several additional lines of the vertical blanking interval, which may contain user data, are recorded by the composite digital format which are not recorded in the D-1 format.

Audio Sampling:

Audio channel sampling of the composite digital format is essentially the same as that of D-1.

TAPE & TAPE FOOTPRINT DIFFERENCES

Physical characteristics of the format footprints on tape is compared in table 4.3. The following are important differences in the tape and tape footprint.

Tape:

1500 Oersted tape is used in the DCR format instead of the 850 Oersted used for D-1. This higher coercivity provides a 6 dB increase in signal output from the tape. The result is that off-tape signal to noise ratio is significantly improved which improves the basic system error rate.

Also, the improved off tape signal allows recording of shorter wavelengths with a resulting increase in data packing density for the recorder. Increased packing density means more recording minutes on a cassette.

Alternate Azimuth Recording:

Different azimuth is recorded on adjacent helical tracks in the composite digital format. With a relative azimuth difference between two tracks of 30°, adjacent track signals are naturally rejected. With no loss of tracking reliability this allows a narrow track to track pitch with zero guard band between tracks. This is one of the important factors that allowed increasing the composite digital format record/play time on a cassette.

Channel Coding:

Miller-squared channel encoding is used by the composite digital format instead of randomized NRZ. Miller-squared coding is DC free and has little low frequency content. Therefore a new recording will very effectively overwrite an old recording without the need for separate erase heads.

Because a Miller-square encoded signal lacks low frequency information in the code it also lacks long wavelengths on tape. Therefore the adjacent track crosstalk rejection for a signal on an adjacent track with different azimuth is enhanced.

Audio Sector Placement:

The composite digital format places the audio sectors at either end of the helical track as opposed to in the center. The primary reason for putting the audio sectors at the ends of the helical track is that data recovery during picture in shuttle and variable speed is improved.

ERROR CORRECTION/CONCEALMENT PARAMETER DIFFERENCES

In Table 4.4 the error correction and concealment parameters are compared.

For the composite digital format the Reed-Solomon code has the same Galois field generator polynomial as D-1 plus additional code generator polynomial factors for increased protection. Code size of D-1 and the composite digital format differ because each was chosen to match the particular number of bytes recorded per sector.

Another difference is that the composite digital format provides additional protection of the sync block ID by the inner code. This could not be implemented on the D-1 format because the choice of randomization required independent decoding of the ID.

TRANSPORT

Tape Wrap Angle:

One design of a transport that could be used for a composite digital recorder is shown in Figure 4.1. The most notable difference between it and a D-1 transport is a reduction of active scanner wrap angle to 180 degrees. One hundred eighty degrees is a common configuration and permits the adoption of proven threading techniques, in addition to allowing a 2-channel system.

Control Track Location:

Another difference is the location of the control track and longitudinal audio heads. These have been moved 19mm closer to the scanner to improve the control pulse to video tolerance and to permit the design of more compact transport mechanisms.

SECTION 5   SUMMARY

The preceding discussion has described, applied and compared the composite digital format. The capabilities and benefits of this format will provide the foundation for the next generation of general purpose recorders.

Economic considerations have been an important element in the formulation of the format and its implication upon the resulting recorder hardware. Concern for both acquisition and re-occurring costs of the resulting equipment and media guided the format design process.

The capability for all "tricks" and operational requirements that will be needed over the next decade is an integral part of the format design. Existing systems with analog recorders will smoothly absorb the composite digital recorders with immediate improvement in picture and sound quality over one or many recording generations.

The composite digital format provides a realistic and efficient transition from analog to digital recording for the majority of professional recording applications now and in the future.

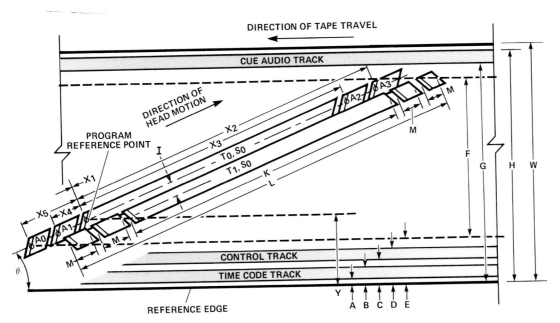

NOTES: 1. A₀, A₁, A₂, A₃ ARE AUDIO SECTORS.
2. T₀, T₁ ARE TRACK NUMBERS, S₀ IS SEGMENT NUMBER (TYPICAL).

Fig. 3.1. Location and Dimensions of Recorded Tracks
(Tape Viewed From Magnetic Coating Side).

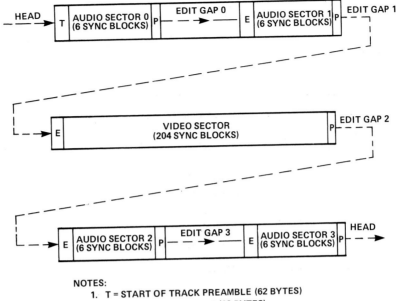

NOTES:
1. T = START OF TRACK PREAMBLE (62 BYTES)
2. E = EDIT GAP PREAMBLE (28 BYTES)
3. P = POSTAMBLE (6 BYTES)
4. SYNC BLOCK = 190 BYTES
5. EDIT GAP = 156 BYTES NOMINAL

Fig. 3.2. Helical Track Arrangement.

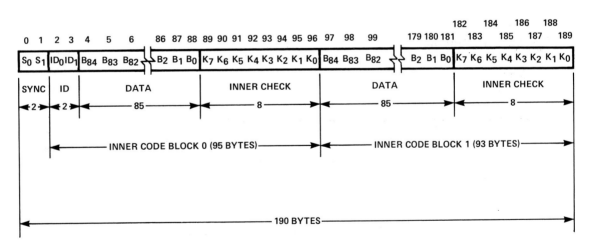

Fig. 3.3. Sync Block Form.

A ARRANGEMENT

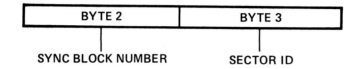

B SECTOR/TRACK ID FOR AUDIO SYNC BLOCKS

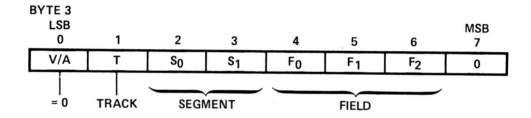

C SECTOR/TRACK ID FOR VIDEO SYNC BLOCKS

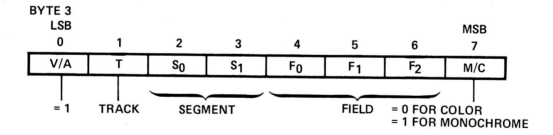

Fig. 3.4. Sync Block Identification Pattern.

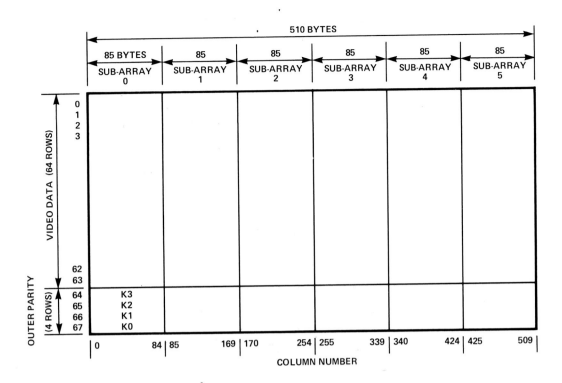

Fig. 3.5. Video Sector Product Block Array

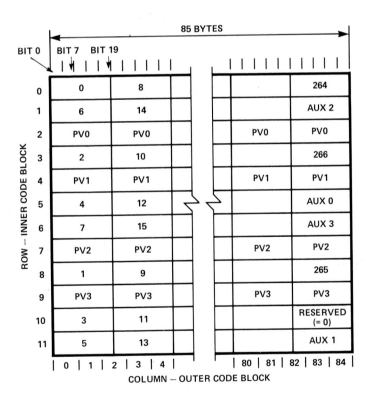

NOTES:

1. NUMERIC TABLE ENTRIES ARE AUDIO SAMPLE NUMBER.
2. SAMPLE 266 IS UNOCCUPIED ONE BLOCK EVERY 5 FIELDS.

Fig. 3.6. Audio Sector Product Block Array.

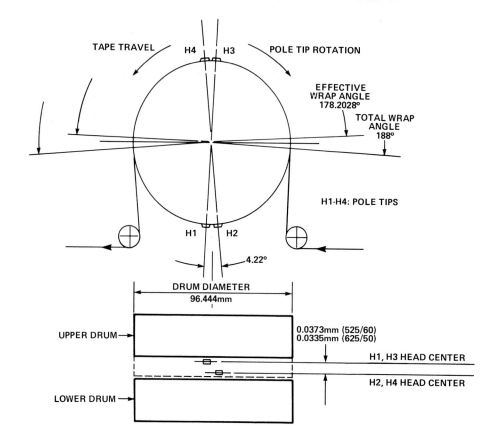

Fig. 3.7. Scanner Configuration

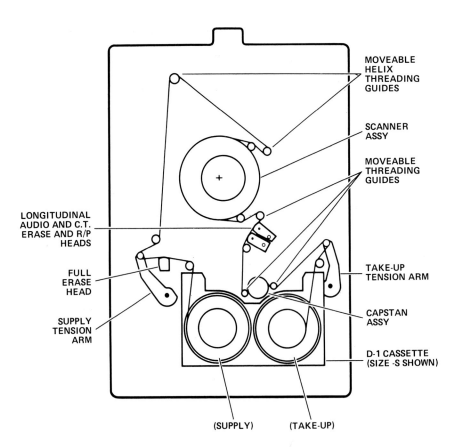

Fig. 4.1. Transport Layout.

Table 3.1. Encoding Parameter Values

AUDIO

==

Sampling Frequency	48.00 kHz locked to horizontal frequency
Form of Sampling	Two's complement, linear PCM 16 bits per sample Provision in format for 20 bits

----

VIDEO

==

Subcarrier Frequency (Fsc)	3.579455MHz
No. of Samples/Full Line	910
Sampling Structure	Orthogonal
Sampling Frequency	4 times subcarrier
Sampling Phase	(I, Q)
Form of Coding	Uniformly Quantized PCM, 8 bits per sample
Number of Samples per Digital Active Line	768

Table 3.2. Record Location & Dimensions for NTSC

	Dimension	Nominal (mm)	Tolerance (mm)
A	Time code track, lower edge	0.2	+/- 0.1
B	Time code track, upper edge	0.7	+/- 0.1
C	Control track, lower edge	1.0	+/- 0.1
D	Control track, upper edge	1.5	+/- 0.05
E	Program area, lower edge	1.807	Derived
F	Program area, width	16.1	Derived
G	Audio cue track, lower edge	18.2	+/- 0.1
H	Audio cue track, upper edge	18.9	+/- 0.1
I	Helical track, pitch	0.0391	Ref.
K	Video sector, length	132.49	Derived
L	Helical track, total length	150.78	Derived
M	Audio sector, length	4.01	Derived
P	Audio/time code/control track	107.66	+/- 0.3
R	Recording tolerance	N/A	+/- 0.1
Q	Track angle	$6.1296°$	Basic
W	Tape width	19.01	+/- 0.015
X_1	Location of video sector	0.0	+/- 0.1
X_2	Location of start of audio sector 3	137.57	+/- 0.1
X_3	Location of start of audio sector 2	133.03	+/- 0.1
X_4	Location of start of audio sector 1	4.54	+/- 0.1
X_5	Location of start of audio sector 0	9.08	+/- 0.1
Y	Program area reference	2.80	Basic
$Alpha_0$:	Azimuth angle, track 0	$+14.97°$	$+/- 0.17°$
$Alpha_1$:	Azimuth angle, track 1	$-15.03°$	$+/- 0.17°$

Table 3.3. Scanner Parameters

Parameters	NTSC
Scanner rotation speed (rev/sec)	90/1.001
Number of tracks per rotation	4
Actual drum upper diameter (mm)	96.444 +/- 0.005
Actual drum lower diameter (mm)	96.429 +/- 0.005
Center span tension (N)	0.7 +/- 0.1
Helix angle (degrees)	6.1592
Effective wrap angle (degrees)	178.2028
Scanner circumferential speed (m/sec)	27.3
H_1, H_3 overwrap, leading (degrees)	5.0
H_1, H_3 overwrap, trailing (degrees)	4.8
Angular relationship H_1--H_2 (degrees)	4.22
Angular relationship H_3--H_4 (degrees)	4.22
Angular relationship H_1--H_3 (degrees)	180.00
Vertical displacement H_1--H_2 (mm)	0.0373
Vertical displacement H_3--H_4 (mm)	0.0373
Maximum tip projection (micrometers)	50

Table 4.1. General

	D-1	Composite
Primary Application	Digital component environment	Analog composite environment
Basic Requirements	Specified by EBU & SMPTE User Groups	Followed user group requirements with some additions
Primary I/O	4:2:2 Digital Video 4 Chan. AES Digital Audio	Composite analog video 4 Chan. analog audio
Cassette	New 19mm Cassettes Small, Medium & Large	Same 19mm cassettes as D-1
Play Time D-1.S D-1.M D-1.L	16 micrometer tape 11 min. 34 min. 76 min. (94 min w/13um)	13 micrometer tape 32 min. 94 min. 208 min.

Table 4.2. Data Sampling Parameters

	D-1	Composite
Video		
Sampling Rate	13.5 + 6.75 + 6.75 = 27MHz	4Fsc = 14.31818 MHz
Resolution	8 bits	8 bits
Number of Recorded TV Lines/Field	250	255
Audio		
Sampling Rate	48 KHz	
Resolution	16 to 20 bits	
Ancillary Data	Defined by AES Serial Interface	

Table 4.3. Tape Footprint

	D-1	DCR
Tape		
Width	19mm	19mm
Thickness	13um and 16um	13um Standard
Coating	850 Oe	1500 Oe
Helical Record		
Wavelength	0.9 um	0.85 um
Tracks/Field	10	6
Track Length	170 mm	150 mm
Track Pitch	45 um	39 um
Effective Track Width	35 - 40 um	39 um
Azimuth	0°	±15°
Audio	4 Sectors in center	2 Sectors either end
Channel Code	Randomized NRZ	Miller Squared
Control Track	Pulse Doublet Servo, Video Frame, Color Frame, Audio Frame	Pulse Doublet Servo, Video Frame, Color Frame
Time Code	Single/Double Unresolved	Standard Single

Table 4.4. Error Correction/Concealment Parameters

	D-1	DCR
Error Management	Error detection, correction & concealment	
ECC Structure	Product Code (Inner & outer)	
Correction Code	Reed - Solomon	
Video		
Inner Code	(64, 60) GF (256)	(95, 87) GF (256)
IC Correction	1 Correction & Detection	3 Correction & Detection
Outer Code	(32, 30) GF (256)	(68, 64) GF (256)
OC Correction	2 Erasure Corrections	4 Erasure Corrections
Product Code Curst Correct	$\leq$ 1200 Bytes	$\leq$ 2040 Bytes
Audio		
Inner Code	Same as Video	Same as Video
Outer Code	(10, 7) GF (16)	(12, 8) GF (256)
OC Correction	3 Erasure Corrections	4 Erasure Corrections
Product Code	$\leq$ 180 Bytes	$\leq$ 340 Bytes
Shuffling	Address permutation based on modulo N congruences	
Sector Structure	Edit Gap - Preamble (with extra Sync & ID) n Sync Blocks - Postamble	
Sync Block Structure	Sync - Block ID - Data - Check Characters	
Sync	30F5 (Hex)	
Block ID		
Size	4 Bytes	2 Bytes
Protection	Self-Contained	Protected by Inner Code

Edwin W. Engberg joined Ampex in 1965 and contributed to the development of electron beam and laser recorders for data and imagery. He was on the development team of ESS-1, which won an Emmy in 1981. After three years as the audio tape marketing manager for Ampex's Magnetic Tape Div., he rejoined the Audio-Video Systems Div. in 1985 as product line manager for 19mm digital video recorders.

Video Research for the Human Viewer

J. K. Clemens
RCA Laboratories
Princeton, New Jersey

Video research for the human viewer begins with research on how a human being perceives a moving image. The research includes studies on how the eye processes images with different parameters such as brightness, contrast, distance from the eye and resolution. Coupled with these studies is research on how the human brain processes more complex stimuli. For example some questions that can be asked are: What is the effect of the visual persistence?; How is motion treated?; And what is the effect of size and brightness on how the the brain treats the information about the image.

The results of this research are crucially important to questions such as: What is the best picture that can be displayed through a channel of a given restricted bandwidth?; What is the best display for a given cost?; What is the optimum frame rate? And how is motion best treated?.

Research at the RCA Laboratories includes the total system which brings the video image to the human viewer. The goal is to develop metrics which can be used to compare system tradeoffs. The research includes hardware and software system simulation as well as study of the human viewer. The research programs and some of the pertinent results will be presented.

I.   INTRODUCTION

Television research has been continuous at RCA Laboratories for 44 years. Many significant results have come from this research and development, perhaps the most significant was the development of the NTSC color system. In all of the large scale programs, the entire television system has been considered. Others sometimes think of the total TV system as light-to-light, i.e. the light input into the camera through the system to the light that the display produces. We, however, do not consider light-to-light to be the total system. Just as there is no sound when a tree falls in the woods if no one is there to hear it, the display does not make an image unless someone perceives it. Figure 1 depicts the entire system as we describe it. The total system includes the human viewer, which is described by a vision model. The vision model includes the human eye, the connections to the brain with all associated filters and the processing that the brain performs. Serious errors and false economies can result if the vision model is left out of the system.

II.   THE VISION MODEL

A. IN THE BEGINNING

The early work on which the television system was based concentrated on what the eye could resolve and television picture quality was judged by its resolution, usually implied by the number of scan lines. Work by Schade (1,2) and others extended the concepts to measurements of the MTF and the threshold characteristics of the eye. This allowed comparisons of the image to include the gray scale and the granularity of noise as well as the sharpness and definition.

The early work also included an operational understanding of an interaction of the eye and the brain in the perception of the image because of the problems encountered with flicker. The field rate was determined by experiment. There was also an understanding that flicker and brightness were related. There was no real understanding of the eye brain interaction to the extent that models could be formulated and research results predicted.

Today we know of the importance of the eye brain interaction in perception. This is evidenced in the multiple channel model of the visual system and in the temporal response and to a certain extent the interaction of these models.

B. THE MULTIPLE CHANNEL MODEL

The evidence for a multiple channel model of the human vision was first indicated by Campbell and Robson (3) in 1968. A quantitative model was developed by Wilson and Bergen in 1979 (4). The model is based on the perceived response of a sinusoidal stimulus, shown here in Figure 2. The sensitivity of the visual system (S) is a function of frequency(F), measured in cycles per degree. The sensitivity falls off at high frequencies as one would expect, the surprise is that it also falls off at low frequencies. The peak of the response varies somewhat with individuals, but is in the range of 4-6 cycles per degree.

The significant discovery for the multichannel model is that the human visual response can be separated into independent channels or bands as shown in Figure 2. These bands are independent in that the image components or noise in one band do not mask or interfere with the image components in another band. Figure 3 shows what an image looks like when it is separated into the band pass channels that are indicated in Figure 2. It seems obvious that these separate channels have very different characters to them and it may not be surprising that they can be treated quite differently in a television system. This fact leads to some very interesting applications, which will be mentioned later.

C. THE TEMPORAL MODEL

The time response of the human vision system is very important to the total television system. Certain aspects are well understood and are included in optimal television display systems. Other aspects are at the outer limits of understanding and are presently under research at RCA and other laboratories.

The time response of the visual system as it relates to sampled systems such as movies and television when displaying a stationary image can be considered to be a linear filter. If the sample rate is fast enough the filter will smooth out the samples; however, if the sample rate is too slow the samples will become apparent and annoying (e.g. a shutter rate of 30 Hz.)(5). Modern television systems and displays are designed to make use of this fact using high frame, field and shutter rates to reduce artifacts.

However, television is based on moving pictures. A great deal of research has been done to try to determine a model for how the human vision system determines motion, of particular interest is how motion is inferred in a sampled system. Many of the models assume some form of feature matching (6,7,8). In this model the visual system is assumed to find the same feature in succeeding frames and to determine the velocity and direction from the distance moved.

At RCA Laboratories another model has been proposed by Adelson and Bergen (9) which matches the measured human vision system very well for small displacements between frames. This model assumes a receptive field with a spatial and a temporal response. The receptive field is a combination of simple cells in the visual cortex which operate as a linear filter. This receptive field has a spatiotemporal impulse response and the vision system can determine the motion without feature comparison. Use of this model of the temporal response of the human visual system has lead to some interesting applications which will be described later.

Ongoing research studies at RCA Laboratories suggest that there is a band pass analysis in human perception of motion which is similar to the multichannel band pass spatial analysis in human perception. Evidence of this model can be experienced in the viewing of a videotape specifically generated for this purpose. The motion model has already produced surprising results toward the problem of frame conversion, which will described later.

III.   APPLICATION OF THE HUMAN VISION MODEL IN TELEVISION
        SYSTEMS

A. JND MODEL

The JND or Just Noticeable Difference is a well-known subjective metric. When comparing two subjective measures (e.g. Which warm plate feels warmer?), one JND is defined as that difference which causes a correct decision 75% of the time. Three JND's difference causes a correct decision 99% of the time. Ten to twenty JND's is therefore a very large difference.

The JND subjective metric has been applied to perception of an image by RCA researchers (10). The JND metric can be used to compare television systems. For example, one system may produce an image which is 10 JND's better than the image produced by another system.

In Figure 4 is depicted an image from a given television system which is broken into 5 multiple channels. The response in each channel can be calculated or measured. Then the calculation or measurement can be performed on a different television system. The difference between the responses for the two systems can be represented in JND's. Since each of the channels is independent, the difference between the two systems in JND's can be found by adding the responses of the 5 different channels.

We have found this method of total system comparison, which includes the model of the vision system, to be very powerful. It has been used to study many different system parameters. One such study was the comparison of HDTV systems which was presented at the SMPTE conference in 1984 (11). Another study which was performed was the pitch requirements for CRT character displays. The inclusion of the human viewer in the system study allowed us to find the optimum in cost performance. Another study, not yet complete, is in the optimization of peaking performance for TV receivers.

B. THE TEMPORAL MODEL

Perhaps the most interesting application of the temporal model of human vision today is in frame conversion. The problem occurs in conversion between standards, in displays and in data compression systems. The most difficult part of the problem is shown in Figure 5 where the conversion is from a lower frame rate to a higher frame rate. The problem is what information to put on a frame which must be "manufactured".

It should be noted that this problem is not a difficult one for the human. It is true that if the frame rate is too slow, flickering and jerky motion will be noticed and will be annoying. However, if two successive still frames, taken not too far apart in time, were shown to a human viewer he or she could easily describe in words how the objects moved or twisted between the two frames. In addition if the person was a artist, he or she could generate the missing frame or frames.

In Figure 6 the problem of generating a missing frame is broken down into its elemental parts. Here we have just 2 frames and we wish to generate one in the middle. One type of frame conversion equipment uses the repeat frame method. This is very inexpensive but causes jerky motion at low frame rates. Another common method of frame conversion uses frame averaging where the middle frame is an average of the two end frames. In this case the motion is smooth but a double image appears in rapid motion sequences. The correct way, of course, is to analyze the motion implied by the two frames and to put the correct frame in the middle much as the artist described above would do it. The model of the human perception of motion described earlier has permitted RCA researchers to generate the missing frame in an automatic way. That is to say, no characteristics of the image need to be described to the algorithm. The algorithm creates the missing frame by analyzing the two end frames without any information describing the image added by the researchers. The results are relatively spectacular.

A video tape has been prepared by T. Adelson and J. Bergen to demonstrate the motion interpolation algorithm. Two black and white still images of a person taken about 1/2 of a second apart were used as the end frames. In order to magnify the results, 5 frames were motion interpolated between the end frames. The resultant is then played at 15 frames per second. The tape includes frame repeat, frame averaging and motion interpolation. The motion interpolation is clearly far superior.

IV.   CONCLUSIONS AND FUTURE RESEARCH

The importance of including the human viewer in the total television system is not in dispute. The display is not an image until it is perceived. Early television systems took into account certain properties of the eye and considerations of flicker. Today our understanding of the human visual system allows a much more cost effective design of television systems. Already the multichannel model has allowed RCA to design display products which are more cost competitive than other products of the same perceived quality. It is also being used to evaluate EDTV and HDTV systems. The temporal model of motion interpretation is just beginning to be used in system design.

Present research in motion perception at RCA suggests a band pass analysis of motion similar to the spatial band pass analysis. This research leads to the development of a temporal JND model. The temporal JND model is not completed, but we expect the results of the model to be as significant to the development of television systems as the spatial JND model.

ACKNOWLEDGMENTS

The author has for the most part reported on the work of others at RCA Laboratories. The references cited refer to only a fraction of the work that has been done in video research for the human viewer.

The author wishes to especially thank Peter J. Burt for his assistance and the members of Advanced Image Processing Research of which Peter is the Head for their contributions.

REFERENCES

1. O.H. Schade, Sr., "Electro-Optical Characteristics of Television Systems," published in *RCA Review*, Vol. 9, in four parts, "Part I: Characteristics of Vision and Visual Systems," p. 5, March 1948; "Part II: Electro-Optical Specifications for Television Systems," p. 245, June 1948; "Part III: Electro-Optical Characteristics of Camera Systems," p. 490, Sept. 1948; and "Part IV: Correlation and Evaluation of Electro-Optical Characteristics of Imaging Systems," p. 653, Dec. 1948.

2. O.H. Schade, Sr., "Optical and Photoelectric Analog of the Eye," *J. Opt. Soc. Amer.*, Vol. 46, p. 721, Sept. 1956.

3. F.W. Campbell and J.G. Robson, "Application of Fourier Analysis to the Visibility of Gatings", *J. Physiol.* 197, 551-566, 1968.

4. H.R. Wilson and J.R. Bergen, "A Four Mechanism Model for Threshold Spatial Vision," *Vision Res.* 19, 19-33, 1979.

5. A.B. Watson, A. Ahumada, Jr., and J.E. Farrell, "The Window of Visibility: a Psychophysical Theory of Fidelity in Time-Sampled Visual Motion Displays," NASA Tech. Paper TP-2211, 1983.

6. S. Ullman, "The Interpretation of Visual Motion (MIT U. Press, Cambridge, Mass., 1979).

7. S.M. Anstis, "The Perception of Apparent Movement," *Phil. Trans. R. Soc. London Ser. B* 290, 153-168, 1980.

8. S.M. Anstis, "Apparent Movement" in Handbook of Sensory Physiology, Vol. VIII, Perception, R. Held, H.W. Leibowitz, and H.-L. Teuber, eds. (Springer-Verlag, New York, 1977).

9. E.H. Adelson and J.R. Bergen, "Spatiotemporal Energy Models for the Perception of Motion", *Journal of the Optical Society of America A*, Vol. 2, p. 284, February 1985.

10. C.R. Carlson and R.W. Cohen, "A Simple Psychophysical Model for Predicting the Visibility of Displayed Information", Proceeding of the Society for Information Display 21, pp. 229-246 (1980).

11. C.R. Carlson and J.R. Bergen, "Perceptual Considerations for High-Definition Television Systems, *SMPTE Journal*, Volume 93, No. 12, December 1984.

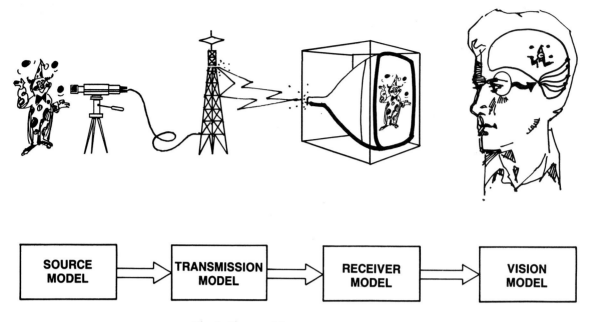

Fig. 1. Human vision: an engineering model.

MODELING THE VISUAL SYSTEM

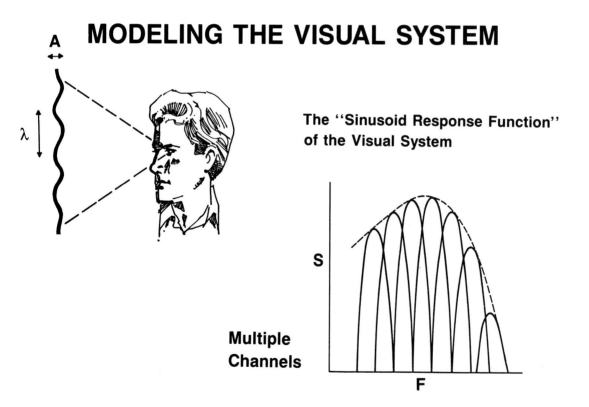

The "Sinusoid Response Function" of the Visual System

Multiple Channels

Fig. 2. Modeling the visual system.

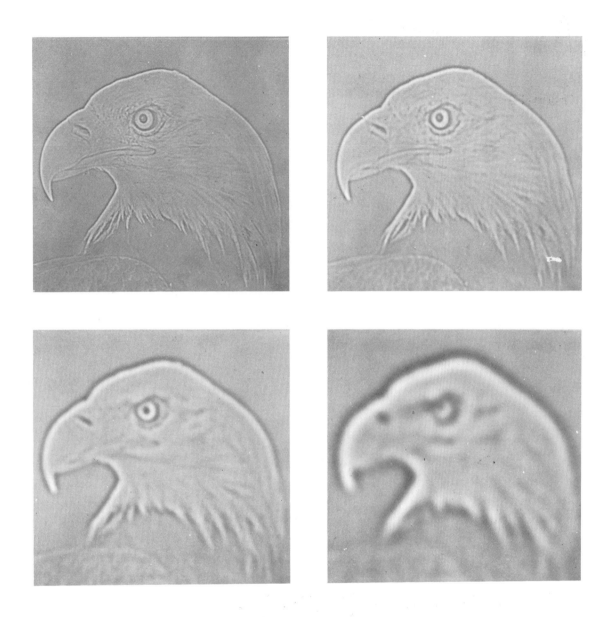

Fig. 3. Band-pass images.

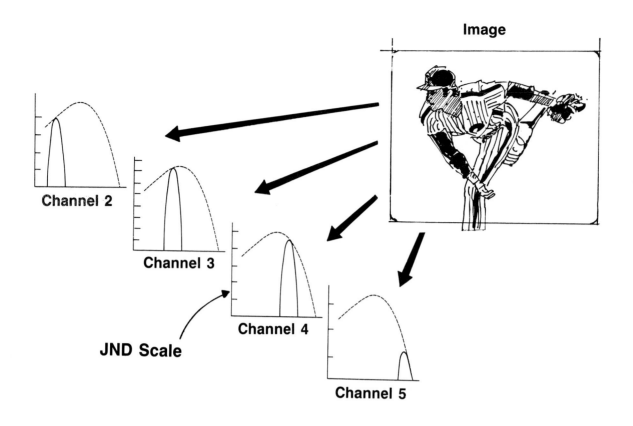

Fig. 4. The JND model.

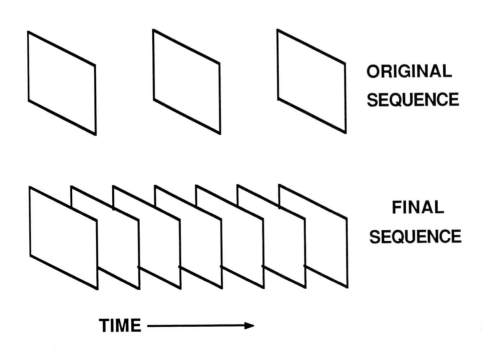

Fig. 5. Frame—rate conversion.

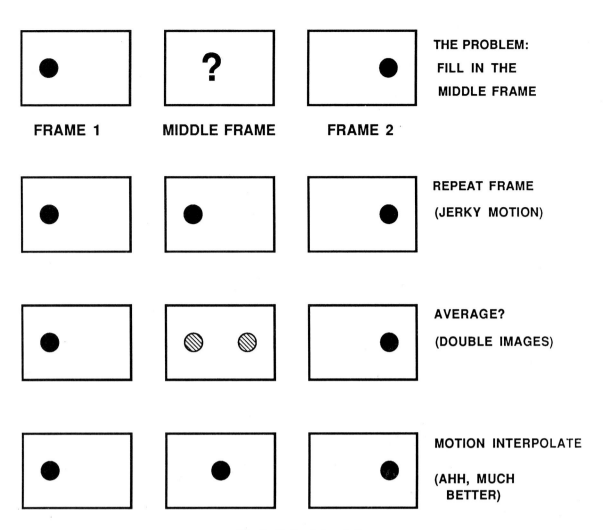

Fig. 6. Motion interpolation.

Jon Clemens received a Ph.D. degree from MIT in September 1965. His thesis work was on optical character recognition. He joined RCA Laboratories in 1965 to work in consumer electronics research. Most of his technical work has been concentrated on consumer video products. In 1983 he was appointed staff vice president, consumer electronics research at RCA Laboratories, and he presently fills that position. In this capacity he is in charge of consumer electronics research for the RCA Corp. He has three research laboratories reporting to him: Television Research Laboratory, Digital Products Research Laboratory, and Information Research Laboratory. Jon has received three RCA Laboratories Outstanding Achievement Awards and the David Sarnoff Award for Outstanding Technical Achievement, RCA's highest technical honor. He was a corecipient of the 1980 Eduard Rhine Prize, presented by the Eduard Rhine Foundation of West Germany, for major improvements in television systems and a corecipient of the IEEE Vladimir K. Zworykin Award in 1983.

Technical Research and Development for the Growth of Tomorrow's Broadcasting Business

Masahiko Ohkawa
The NHK Science & Technical Research Laboratories
Tokyo, Japan

Television broadcasting has rapidly developed and now, technically it is progressed nearly as far as it can go, and it is groping for further development toward the establishment of an advanced television system. This paper describes the author's views on trends in broadcast technology which need to be studied and researched for the further growth of broadcasting in the 21st century. Television broadcasting can provide a great amount of information quickly, and over a wide area. It is indispensable for our everyday lives as a medium which gives us at very little cost, thought-provoking and entertainment programs with high-quality pictures and sound. In the second half of the 20th century, the rapid spread of television has transformed man's life and culture, and it has grown into a telecommunications media which has a great influence on society. This outstanding growth of television has had, as its basis, farsighted technical research. For further growth in broadcasting, it should meet viewers' needs and expectations in the high-level information society of the future.

In such a society, electronics, telecommunications and the information media will be used in a sophisticated way both for social life and home life. Catering to the needs of the individual, diversification, and internationalization will be more common due to the rise in living standards and an increase in leisure. The role of broadcasting will be to provide high-level, diversified, internationalized services with the aid of more efficient management and operation of the broadcasting business.

Research and development are closely related to the running of broadcasting in the future, and to support management. Research must be conducted within a broadcasting organization for the development and realization of its engineering means, constantly providing and assigning talented young researchers in accordance with management policy and the needs of the operational side. From this point of view, the research and development to be achieved are classified under two categories: (1) those for new services and (2) those for the efficient carrying out of broadcasting, making full use of the latest frontiers of technology.

(1) RESEARCH AND DEVELOPMENT FOR NEW SERVICES

Research and development for new services is indispensable for the expansion of the broadcasting business. HDTV, recently named Hi-Vision, stereoscopic television, Integrated Service Digital Broadcasting (or I S D B) are considered important services related to the visual side of TV. To bring new services into the home, it is necessary to exploit the transmission media which distribute these new services.

Hi-Vision

Hi-Vision is an entirely new television concept which fulfils viewers' demands for a high-quality picture. It is based on thorough research and a study of the physiology of how we use our eyes, taking into account psychological factors as well. Latest technological methods have been extensively used. Hi-Vision conveys five times the amount of information received by conventional television, with excellent picture quality. It has a sense of presence and reality, and has a great impact on the viewer. There is no doubt it will create a new visual culture and it holds out promise for a more advanced information society.

Almost all the equipment from image pickup to transmission and reception has been developed and it is in practical use. The most important thing to do now for the realization of Hi-Vision broadcasting is the development of low-cost receivers; flat panel displays which can be used in the home and high-sensitivity image pickup devices must be further studied and developed; in fact, these are actually being developed.

Low cost is most important for home-use receivers and the development of LSIs is being carried out for the receivers including decoders for the MUSE signal which can be transmitted on one channel allocated to satellite broadcasting.

The plasma display which is being researched is most appropriate, for large-size and high-quality Hi-Vision receivers among various other panel displays. Our research has been stressed on achieving higher emission efficiency by using an 8-inch colour plasma display. It is now the time to start developing larger panel to find a solution of manufacturing problems.

It is thought that it will take 7 to 10 years for a Hi-Vision panel display to be available in the home. Before its realization, projection-type displays will be used as a temporary measure for the time being. The largest Hi-Vision display ever developed is a 400-inch rear-projection type which is still being used at the Cosmic Hall in Tsukuba Academic City. It was made for EXPO 85 under the direction of NHK. It has twelve 10-inch projection CRTs and could very well be the prototype for Hi-Vision theaters of the future.

Another important task is to develop high-sensitivity image pickup devices. The bandwidth for Hi-Vision is five times as wide as conventional television. Naturally, it produces higher noise and needs higher sensitivity. In addition, as the viewing angle is wider, the iris has to be stopped down to give the same sharpness of image as in conventional television. High sensitivity is essential for a Hi-Vision camera for it to be used under the same conditions as in conventional television. The NHK Science & Technical Research Laboratories have been working on this, and are expecting good results from it.

Stereoscopic Television

In general, research and development of Hi-Vision equipment has come to an end, and it is now time to make our first step toward advanced broadcasting systems of the future. "Stereoscopic television" is an example of technology, one step beyond Hi-Vision. Stereoscopic television will add a sense of depth to Hi-Vision. It gives a more natural and realistic image. For many years, various types of stereoscopic pictures and vision through the use of special glasses have been developed. They have been demonstrated to the public, but they are not satisfactory and have failed because of their exaggerated and unnatural images and because of the fatigue induced in the viewer. Stereoscopic television will never be attained unless and until Hi-Vision technology is taken into consideration for this new development.

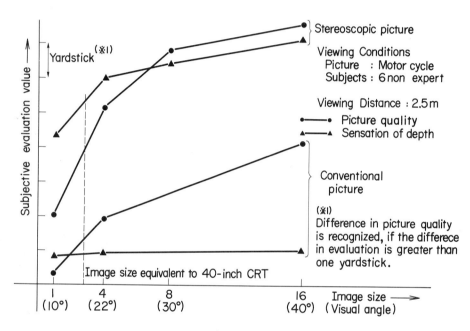

Fig. 1. Comparison of picture quality between
stereoscopic and conventional pictures.

In order to develop a stereoscopic television system which matches the visual responses of human beings without using any glasses, basic study is necessary on the mechanism of the eyes' physiology, and on psychological aspects as well. Fig. 1 is one result of our study. It shows that a stereoscopic image has obtained a higher evaluation than two-dimensional images. Our effort will be concentrated on finding a suitable method for reproducing a stereoscopic image as real as that seen by human eyes, through various research projects.

I S D B

"I S D B" stands for Integrated Service Digital Broadcasting. I S D B is considered to be one of the broadcasting media, meeting the needs of diversification in the future

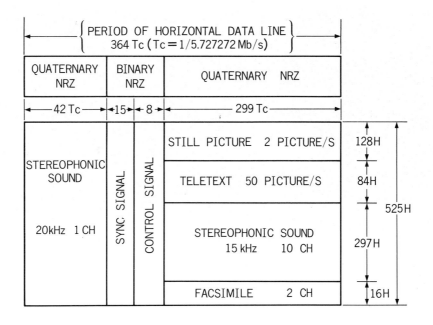

Fig. 2. An example of signal construction of ISDB.

society. It can be used for today's television system, Hi-Vision, sound broadcasting, teletext, facsimile, still pictures and data broadcasting through one transmission carrier digitally and integratedly.

Fig. 2 shows one example of the signal construction of ISDB, in which television transmission is not included. The signal construction is based on the conventional television signal, but effective utilization is available with a combination of sound and still pictures, data and facsimile and so on. ISDB is the synthetic and final broadcast system which corresponds to a certain extent with ISDN in the field of telecommunications. The signal format will be determined taking into account ease of operation, integration, and individualization of the receiving terminals.

(2) RESEARCH AND DEVELOPMENT FOR EFFICIENT BROADCASTING OPERATIONS

Efficient management and operation of broadcasting is always important both for new broadcasting services, meeting the social needs of the future and for current broadcasting services.

The following are the research and development considered from technical points of view concerning program production, news and information gathering and sound and picture data storage.

Program Production Technology

It is believed that two types of programs will be in demand as a result of trends in society in the future; one will be news and impressive programs for mass audiences and the other is diversified programs for a small number of special, specific audiences. As for this second type, it will be necessary to be efficient and inexpensive program production through the effective introduction of computer technology, recognition and artificial intelligence.

Through the introduction of office automation, sound and letter recognition, and sound synthesis, for example, it is quite probable that there will be development of automatic writing of manuscripts, electronics scenarios, automatic promptors, automatic focussing of specified objects and automatic tracking cameras.

As the processig speed of computers is increasing rapidly year after year, development of various expert systems for television program production will be possible, and automatic program production will be common in the 21st century. Automatic translation of programs will become an important subject to be resolved when there is an increase in international exchange of programs and program production cooperation.

News and Information Gathering Technology

In order to gather news and general information material and to transmit it accurately, quickly and with high-quality, the equipment must be extremely mobile and simple to operate.

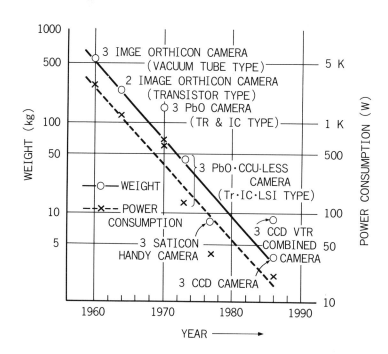

Fig. 3. Weight and power consumption of broadcast-use cameras changed with year.

It must also be compact, light in weight, stable and reliable. It will not be very long before it is considered unnecessary for specialists to be present on location.

Fig. 3 shows the change in weight and power consumption of television cameras from the past to the present. The weight and power consumption decrease exponentially as the years go by. Technology is progressing very fast, and the appearance of a very small television camera for broadcast use is soon at hand.

Recording is another important technology for information gathering. At present, research in perpendicular magnetic recording technology is making progress so that by using this technology, colour moving pictures can be recorded and replayed. This technology will be introduced into the magnetic tape recording field in the future. Its recording density is more than ten times as high as that of conventional tape recording. A pocket-size camera-VTR is possible through the use of higher density tape.

There will be an increasing demand for development of program-production equipment far beyond man's ability. One good example is a camera system which was developed by NHK for picking up Halley's comet. This system has 20,000 times higher sensitivity than a conventional camera, making the best use of signal processing technology and frame memory. When the need arises, such information gathering equipment that can perform what the man never can do will be developed.

Technology of Picture and Sound Data Storage

It is desirable to retrieve the most suitable material as quickly as possible whenever it is required. We can expect development of high-density recording, efficient signal coding and quick retrieval.

Besides perpendicular magnetic recording, great progress is being made in magneto-optical recording, solid-state memory and biomemory. These are also useful for picture storage technology. A magneto-optical disk which was developed by NHK Laboratories can record and replay broadcast-quality moving pictures a number of times. At present, ten minutes can be recorded on two 30cm-diameter disks, but more than ten minutes is possible. The access time for a disk is only 0.5 seconds at most. Quick editing is possible direct from various news materials. Needless to say, it is effective for storage of picture and sound material.

Broadcasting tomorrow will be what we dream of now, and research and development of technology will be the trail-blazers of broadcasting, as they have been in the past. The future of broadcasting is in our hands. Broadcasting will revolve around satellite broadcasting and Hi-Vision, and these will activate the electronics and related industries as well. There will be more frequent and easier international exchange of technology and programs, joint technical research and joint program-production, in accordance with the progress of various technologies. This will surely help to bring about mutual understanding among the peoples of the world, and will contribute to peace on earth.

Masahiko Ohkawa is with the NHK Science & Technical Research Laboratories, Tokyo, Japan, since July 1986 as Director-General. He joined NHK in 1955 after receiving a Bachelor's degree in electrical engineering from Waseda University. From 1960 to 1974 he was involved in taking the lead for the development and operation of the NHK's automatic broadcasting transmission system. Since 1974, he has been active in planning the NHK's broadcasting engineering systems, facilities, and equipment.

Television Engineering Research in the BBC, Today and Tomorrow

Bruce Moffat
British Broadcasting Corp.
United Kingdom

ABSTRACT

Television engineering research in the BBC is intended to ensure that the future engineering needs of the BBC can be met economically and effectively. To that end new broadcasting systems and techniques are innovated and assessed. It is important to strike a balance in this work between current and future requirements, mainly through internal projects, but partly in cooperation with other enterprises. Major examples of relevant work include High Definition Television in conjunction with Digitally Assisted Television, bandwidth compression, subscription systems, digital techniques for studios and transmission, and digital stereophony for television.

INTRODUCTION

Research into television engineering in the British Broadcasting Corporation is centred at Kingswood Warren in Surrey, a 150 year old mansion set in 30 acres of parkland about 20 miles south of the centre of London. The mansion is augmented by several more modern buildings and in all of them, old and new, are to be found excellent facilities for studying the whole area of broadcasting engineering.

Research Department is one of the specialist departments within BBC Engineering Division, the others being concerned with design, equipment, transmission capital projects, architecture, civil engineering, information and training. The Department's principal customers are the Television, Radio and External Services Directorates and Transmission Group. Close liaison is maintained with the other specialist departments and many projects are undertaken in collaboration with equipment manufacturers and higher education establishments.

The role of the Department is to ensure that the future engineering needs of the BBC can be met, economically and effectively. To that end, new broadcasting systems and techniques are innovated and assessed, providing technical advice to the engineering specialist departments and the output directorates, based on theoretical studies and experiments. One of the Department's main outputs is Research Reports. Unnecessary duplication of research and development underway in industry and other broadcasting enterprises is avoided as far as possible.

The staff of about 230 comprises around 120 qualified engineers, physicists, mathematicians, computer scientists and technician engineers, supported by comprehensive technical, administrative, computer and library services. There are eight research sections organized in three groups focused respectively upon studio, transmission and radio frequency topics. Although full facilities for implementing hardware are provided, including the fabrication of service equipment if necessary, there has been a strong trend towards computer software becoming the dominant instrument in all aspects of our work; as a consequence, much more complex studies have become feasible and conventional research measurements and data processing have been speeded up.

The main funding of the Department stems from the three output directorates; Television accounts for about 70% of our income and expenditure. Significant financial contributions and royalties to the BBC from industry are generated as a result of collaborative research agreements and licences.

The rest of this paper touches upon important examples of short, medium and long term research projects being undertaken. It is intended to show how we try to strike a balance in television engineering research between current requirements and future developments, mainly by our own efforts but partly in cooperation with other sections of industry, other broadcasters, and the academic world.

HDTV AND DATV

Work at the BBC on High Definition Television (HDTV) covers a wide range of activities. In addition to the production of superb wider and larger colour pictures [1], directly associated topics such as broadcast frequency planning, transmission compatibility with existing services, multi-channel digital audio, subscription systems (embodying encryption and/or scrambling) [2], and data broadcasting are included. The BBC has made a significant contribution to solving the major problem of bandwidth compression by proposing the promising technique of Digitally Assisted Television (DATV) [3]. Integration of HDTV and DATV with digital studio and transmission developments in general requires careful evolution of existing facilities. All of these topics have been worked on in BBC Research Department, some for several years, and many of the technical, commercial and political problems associated with them will require a great deal of effort through to the twenty-first century.

The 1986 CCIR Plenary Assembly's attempt to reach agreement on a single worldwide HDTV studio production standard foundered mainly because of a lack of consensus on a scanning system and we have therefore continued to study the factors affecting the choice of such a system.

One example of the hardware used in these studies is an HDTV picture storage system which can be operated at several scanning rates. A high line-rate colour picture monitor has recently been used with the storage system in an investigation of the optimum compromise between field rate and spatial resolution in the display context alone; the source pictures were derived from laser-scanned, very high resolution still picture data. The results augmented and supported earlier work [4] directed towards source and display scanning considered together, using a modified television camera and a multi-standard converter. The new equipment provided part of our exhibit at IBC 86 where it aroused considerable interest; for many engineers it was the first time they had seen on one monitor the effects of the trade-offs that can be made between 50 Hz and 60 Hz based HDTV scanning. Interlaced and sequential scanning options were included. It was particularly interesting to note the favourable comments from visitors concerning the quality of the (still) picture display at a field rate greater than 60 Hz.

The choice of an HDTV scanning standard for the studio is less problematical than that for transmission in which a dichotomy immediately becomes apparent in the form of evolutionary and revolutionary approaches (which would ultimately include all-digital transmission). In the former we optimise the compatibility between new and old transmission systems; in the latter we optimise the transmission independently from existing systems. Both are being pursued by the BBC, with greater emphasis being given to the evolutionary approach for the short to medium term. This strategy has been influenced by Eureka Project 95, Compatible High Definition Television, which is getting underway in Europe with Thomson, Philips, Thorn EMI and Bosch as the major participants. Central to the project is the objective of demonstrating a High Definition MAC system receivable by MAC receivers, which will be introduced in Europe for Direct Broadcasting by Satellite during the next year or two, with virtually normal MAC quality and 4:3 aspect ratio. HD-MAC receivers fully exploiting the potential quality of the system would them come on the market in the early 1990s.

DATV is applicable to both evolutionary and revolutionary approaches to HDTV; indeed, there is a widespread belief that DATV is an essential element of successful bandwidth compression for any kind of HDTV broadcasting. It can also be applied to existing terrestrial broadcasting and satellite transmissions. It is a class of hybrid techniques in which bandwidth compressed analogue video signals are transmitted along with digital control and/or additional picture data; a DATV receiver can use the data to reconstruct a final picture of higher quality than that to be seen on a conventional receiver. The degree of quality improvement depends upon the transmitted bit rate that can be made available in the DATV signal. For new but MAC compatible systems a few megabits per second could be made available, sufficient to effect substantial improvements. If the compatibility constraint is removed then higher bit rates and better results are likely, even though spectrum bandwidth restrictions would remain, of course. The bit rate which we could add to present day terrestrial broadcast standard signals is lower, but it is still several hundred kilobits per second, and worthwhile improvements should be possible.

We have identified a line of DATV research which should provide results which are applicable to both of the approaches to HDTV, including improvements to existing television services. Three important elements in this research are motion-adaptive filtering with sub-sampling, motion vector measurement and compensation, and vector quantisation. We included demonstrations of the first two elements in our contributions to IBC 86.

Motion-adaptive filtering and sub-sampling equipment, commissioned in 1985, has been used to test bandwidth compression algorithms [3]. The programmable equipment was operated at 625 lines so that we could use fully developed components and avoid spurious effects. Bandwidth compression factors of about 4:1 have been achieved with very little loss of picture quality except in areas of steady motion.

To cope with moving areas, motion vector compensation has been investigated, initially on our computer image processing system [5]. In the best variant so far investigated, the first step was to compute the two-dimensional Fourier transforms of each area of 64 by 64 pixels in two consecutive image fields. Phase correlation between the resultant spectra followed by inverse transformation gave a correlation surface on which the two axes represented horizontal and vertical image detail velocities; the third orthogonal axis represented the proportion of the image field moving at each velocity. This velocity data was assigned to the correct image areas after measuring the magnitude of the differences between corresponding areas in consecutive image fields. Such operations made it possible to regain picture quality in areas of steady motion. Transmission of a few megabits per second of motion and sampling control data should suffice for HD-MAC applications.

Vector quantisation is a promising example of bit rate reduction techniques which we are examining in the HDTV and DATV context; in principle, it is well suited to broadcasting applications in relation to a wholly digital signal or to the digital part of a DATV signal. Groups of luminance and colour-difference samples are coded together and restricted quantisations of combinations of samples are transmitted. The restrictions are set by adapting the encoding to match the source picture data. Coding is complex but algorithms have been devised and tested which could be applied using real-time hardware. Decoding is relatively simple, and therefore attractive for domestic receivers; it requires a 'look-up' table to convert received codewords into sample value combinations. Effective bit rate reduction ratios of 5:1 have been achieved experimentally.

Two DATV bandwidth reduction techniques applicable to the revolutionary approach to HDTV are what we term slope coding and block adaptive sub-sampling (BASS) [6]. Both of them use variable-rate sampling governed by the amount of spatial detail or temporal activity in the picture. We have examined them in computer simulations and good HDTV quality within about 40% of the original bandwidth was achieved. Motion vector compensation applied to BASS coding may give further compression; investigation continues.

Crucial to HDTV research is the investigation of transmission channel characteristics. For some years we have been assessing the maximum capacity of a DBS channel in the 12 GHz band. We have found that up to 12 MHz of video basebandwidth or between 60 and 100 Mbit/s in digital capacity may be possible. Such values would be reduced by constraining ourselves to a MAC compatible system but they may be attainable in possible new DBS channels in the region of 20 GHz. An equally important aspect of radio frequency work in HDTV research is that of the effects of noise and distortion on the digital component of DATV signals; such work needs to be done but it is already clear that it is essential to have very high quality source signals and to display HDTV pictures at full size before some of the more subtle transmission impairments can be finally assessed.

Reverting to the choice of HDTV scanning for the studio, it is important to note that our motion vector compensation DATV work, aimed chiefly at transmission problems, also has relevance to the studio standard. It can be applied to down-conversion for transmission of a relatively high field rate in the studio, without necessitating complex instrumentation for up-conversion in the domestic receiver. Use of the phase correlation technique to generate the additional fields required in a receiver has been successfully simulated on our computer image processor.

The motion vector compensation technique has also been found in computer simulations to be effective in substantially improving the slow motion replay performance of a conventional 625 line video tape recorder. Similarly, we have been able to show that significant improvements in motion portrayal from replayed motion picture films should be attainable.

Another application of the technique is in interlaced to sequential scanning conversion, an important consideration when choosing the studio and transmission scanning standards for new or improved systems. However, part of our study of such scanning conversion has been aimed at instrumentally simpler methods using novel forms of non-adaptive linear filter and alternative adaptive techniques. These methods have already been applied successfully in frame-grabbing facilities in the Rank Cintel Slide File, developed collaboratively with the BBC as a stills store for studios.

If the field frequency chosen for the HDTV studio is not closely related to the scene lighting power supply frequency, beat frequency flicker can appear in the camera output if arc or fluorescent luminaires are used. For stationary scenes, we have found that the problem can be overcome using a synchronised shutter in the camera or by flicker signal detection and compensation circuits including several fields of storage. However, both of these techniques fail when movement occurs; in the case of an optical shutter the failure is gradual but the electronic technique has a sudden onset of failure. We believe that the flicker problem can be overcome by additional signal processing.

It is difficult and costly to provide all the necessary HDTV source and display equipment, particularly in view of the fact that HDTV cameras and displays cannot yet fully exploit their scanning systems. It will be many years before they do and reaching that goal will depend partly upon new techniques such as CCDs for cameras. Meanwhile, expedient means have to be found to help in avoiding spurious results which could lead to an unsuitable choice of a scanning standard which must last for many decades.

We have almost completed a very high resolution slow-speed slide scanner to give us impeccable source signals. It uses a 3456 element line-array sensor and a micro-positioner; the video signals processing is carried out in the main central computer in the Department (a VAX 11/750).

Recently we have collaborated with Rank Cintel on an alternative way to provide HDTV moving picture sequences; 35 mm colour film is being shot at high frame rates of at least 50 per second for replay on a similarly fast telecine. In this way we intend to attain high spatial and temporal resolution simultaneously and without camera lag.

To approach our HDTV wide aspect ratio display requirements, we have purchased a variable standard CRT projection system and a wide-screen monitor. However, a great deal of useful and reliable research has been done on our high resolution 625 line displays with a 4:3 aspect ratio. We suspended our work on an HDTV laser projection display a couple of years ago but laser technology may eventually become suitable for such purposes.

DIGITAL TELEVISION TRANSMISSION AND ROUTING

Distribution and contribution transmission circuits for video signals using digital techniques have been a major topic in Research Department for 20 years. Nowadays a large number of international organisations are involved in work to define coding standards for the exchange of television signals via digital links. Apart from the CCIR and the EBU, in Europe the COST and RACE organisations are also pursuing developments at bit rates of interest to broadcasters; the BBC is represented on all of these groups. Interest is mainly centred on 34 Mbit/s although higher rates such as 140 Mbit/s are also being considered. The current 34 Mbit/s proposals are based on adaptive forms of DPCM, but we have produced demonstrations which show that it is unlikely that these techniques will be adequate for contribution circuit applications where downstream signal processing, such as chroma-key, is often incurred [7]. We are investigating a possible optimum solution at this bit rate by applying adaptive DPCM techniques to a discrete cosine transform of the input signal. The quality attainable at 34 Mbit/s may well be adequate for signal distribution applications.

The use of optical fibres for digital television transmission is of ever-increasing importance. Relatively recently, it has become apparent that the BBC's medium term needs for television signal routing equipment could be met by exploiting the characteristics of optical fibres. Television Centre in London is a major studio centre in which hundreds of sources have to be routed to hundreds of destinations, and the equipment currently doing this will need to be replaced in the early 1990s. With that application in mind we are pursuing the possibility of using Time-Division Multiplexing combined with Optical Wave-Division Multiplexing (TDM-OWDM) for the transmission of many television signals along a single fibre [8]. By this dual multiplexing technique we hope to avoid the need for optical switches which we are also investigating but which may not become available in the right time scale. The most important optical component which we are currently developing is the wavelength demultiplexer, which consists of a diffraction grating with an appropriately positioned lens, and the optical combiner. To test such devices, a light source of variable wavelength and high spectral purity is needed and so we have constructed an external cavity laser which uses a diffraction grating to control its tuning and selectivity.

We have discussed the TDM-OWDM development with industry and they agree that the idea appears to be feasible. Eventually, fixed wavelength sources will be required and it is confidently expected that suitable feedback lasers will become available. It will soon be possible to obtain the high speed logic elements required for multiplexing and demultiplexing the 2 Gbit/s which will be applied to each laser.

Our colleagues in BBC Television are enthusiastic about this research and we believe that with industrial collaboration, possibly including a European R & D project under the RACE Main Phase programme, it will be possible to meet their time scale. A key advantage of the TDM-OWDM approach to routing is that it could be expanded to accommodate HDTV signals.

DIGITAL STEREOPHONY FOR TERRESTRIAL TELEVISION

Until a few years ago it did not appear practicable to add compatibly a digital stereophonic signal to existing terrestrial television broadcasts. However, thanks to near-instantaneous companding and an ingeniously balanced choice of coding, modulation, signal injection level and carrier frequency we have devised such a system. A standard signal specification for System I has been refined and agreed in discussions with the IBA and receiver manufacturers and in September 1986, just before IBC, it was adopted as a United Kingdom standard.

The system uses 14 bit to 10 bit Near-Instantaneous (NI) companding with error protection and spare signal capacity giving a total bit rate of 728 kbit/s; it has a high degree of commonality with the sound and data multiplex of the EBU MAC Packet family. The modulation is quadrature phase shift keying and the centre of the digital audio spectrum is placed 0.55 MHz above the existing FM sound carrier; the digital signal level is 20 dB below that of the vision carrier, sufficiently low to avoid adjacent channel interference [9].

No firm decision has yet been taken on introducing digital stereophonic broadcasts into BBC Television's programme schedules but plans for implementation have been made and experimental broadcasting continues from London. These broadcasts have included the wedding of the Duke and Duchess of York, orchestral concerts, sport, comedy and pop music. Several dramas have been recorded. All of the main BBC studios are equipped for stereo but it will be a few years before widespread transmission facilities are in place.

Several other countries have expressed interest in the UK system. Sweden, Finland, Norway and Denmark, in which Television System B is used, are adopting it with a suitable spectral placement of the digital carrier. Hong Kong uses System I and is also showing great interest. BBC Research Department has been working closely for the past two to three years with the broadcasters in Hong Kong, Sweden and Finland, to ascertain the suitability of our system in their situations; a successful outcome has been achieved after several series of experiments. Tests with the 728 NI system are also taking place in New Zealand and we are now reasonably confident that it could be applied to Television System L as used in France. The issues involved in setting a European digital stereo standard for terrestrial TV are being vigorously pursued in the EBU.

A crucial factor in the relevant deliberations is the availability of receivers to the consumer electronics market. The BBC is giving advice to receiver manufacturers and Thorn EMI Ferguson and Philips, at least, are developing prototypes. They have in mind the objective of Sweden and Finland to begin TV stereo and dual language broadcasts, respectively, by about the end of 1987. It is worth stressing that the UK system is suitable for the dual language application.

Introducing an addition to an existing system always brings up a wide range of compatibility queries. Two in particular have necessitated a substantial amount of experimental work, namely, potential interference from digital stereo broadcasts in the UK into French television broadcasts, and the degree of compatibility with cable transmission equipment. Lack of evidence on criteria for existing sound protection ratios recommended by the CCIR caused us to conduct extensive tests which confirmed that no interference to French television attributable to the new system would occur. We also undertook experiments to provide advice to UK and Swedish cable television operators on additional protective filtering which they may need to use.

CONCLUSION

In this paper we have outlined several major examples of television engineering research in the BBC, ranging from the longer term but evolutionary issue of HDTV broadcasting to the now shorter term research matter of digital stereophony for terrestrial television broadcasting. Further examples have been presented by means of captioned illustrations. Overall, the intention has been to show that although BBC Research Department is already pursuing programmes which may not come to full fruition in broadcast service before the end of the century, it also pays close attention to matters which are of immediate interest to our colleagues throughout the BBC.

ACKNOWLEDGEMENTS

The author is indebted to many colleagues at Kingswood Warren who have contributed directly or indirectly to this paper, notably Howard Jones, Richard Sanders and Paul Ratliff. He also acknowledges the permission to publish the paper given by the BBC's Director of Engineering.

REFERENCES

1. SANDBANK, C.P. and CHILDS, I., 1985. The Evolution towards High Definition Television. Proc. IEEE, 1985, Vol. 73, No. 4, April 1985, pp. 638-645.

2. EDWARDSON, S.M., 1985. A Conditional Access System for Direct Broadcasting by Satellite. IERE Journal, Vol. 55, No. 11/12, November/December 1985, pp. 377-385.

3. STOREY, R., 1986. HDTV Motion Adaptive Bandwidth Reduction using DATV. IBC 86, IEE Conference Publication No. 268, pp. 167-172.

4. CHILDS, I. and ROBERTS, A., 1984. The Compatibility of HDTV Sources and Displays with Present Day TV Systems. IBC 84, IEE Conference Publication No. 240, pp. 111-115.

5. THOMAS, G.A., 1986. HDTV Bandwidth Reduction by Adaptive Subsampling and Motion Compensation DATV Techniques. 128th SMPTE Technical Conference, New York, October 1986, Preprint No. 128-49.

6. KNEE, M.J., 1986. Bandwidth Compression for HDTV Broadcasting. Proc. of International EURASIP Workshop on Coding of HDTV, Vol. 2, L'Aquila, Italy, November 1986.

7. WELLS, N.D., 1984. Digital Video: YUV Bit Rate Reduction for Broadcasting Applications. IBC 84, IEE Conference Publication No. 240, pp. 338-340.

8. OLIPHANT, A., MARSDEN, R.P. and ZUBRZYCKI, J.T., 1986. An Optical System for Routing Digital Component Video Signals. IBC 86, IEE Conference Publication No. 268, pp. 275-279.

9. ELY, S.R., 1986. Progress and International Aspects of Digital Stereo Sound for Terrestrial Television. IBC 86, IEE Conference Publication No. 268, pp. 138-143.

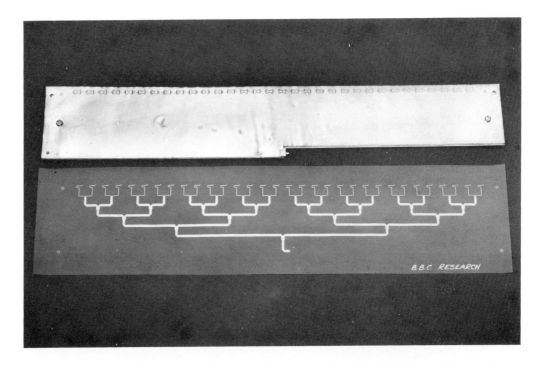

Fig. 1. Unobstrusive flat plate receiving antennas for direct broadcasting by satellite are an attractive alternative to parabolic dishes. They could be attached to a wall and designed as a multi-element array which can squint up at the satellite. In the longer term, phased array techniques may become an economic possibility, offering the facility of beam steering to pick up different satellites without moving the flat plate. The upper part of the photograph shows an experimental array comprising one line of double folded dipoles printed on a Kapton sheet on polystyrene foam above a ground plane; the corresponding dipole feed structure printed on a similar sheet is shown below the line of dipoles. A complete array will comprise several of the linear subarrays.

Fig. 2. Scale modeling techniques have been successfully used for architectural acoustics investigations in the BBC concerned with music studio acoustics and sound insulation. The photograph depicts a recent example of the latter in which the effect of external road traffic noise is being assessed in relation to the new Television Theatre in the Phase Five development of Television Centre in London. The sound source is shown on the left on the model road, the microphone is close to the Theatre wall, right of center, and reflecting surfaces representing neighboring buildings are lower left and right.

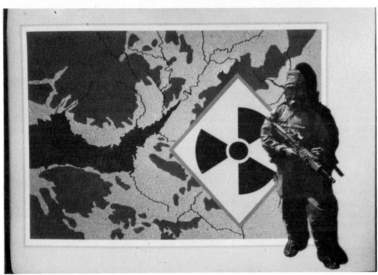

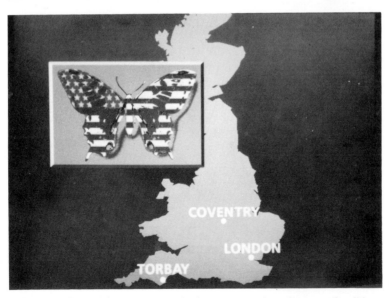

Fig. 3. Slide File is a television stills storage system, conceived as an economic replacement for slide scanners, which has been developed by the BBC in collaboration with Rank Cintel to include "frame grabbing" and to offer the option of being upgraded to Art File for electronic graphics. The Research Dept. has worked closely with BBC Graphic Designers to ensure that Art File is easy to use and offers desirable operational facilities. The photographs show an Art File being assessed in BBC Bristol together with two examples of pictures drawn and composed on the system.

134

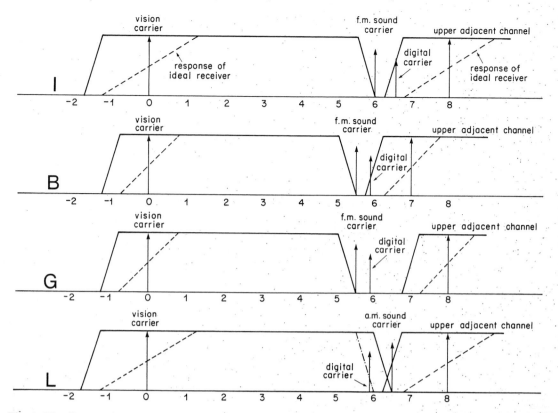

Fig. 4. The diagram shows the spectral positions of the two-channel digital audio carrier, the normal sound carrier, and the vision carrier for Digital Stereo Sound for Terrestrial Television (DSSTTV) for use with Television Systems I, B, G, and L. The digital audio spectrum is about 0.5 MHz wide with the digital carrier level 20 dB lower than the vision carrier for Systems I, B, and G, and 25 dB lower in System L. The System I DSSTTV scheme is now a United Kingdom Standard, and the Nordic countries (Sweden, Finland, Norway, and Denmark) have adopted the schemes depicted for Systems B and G. The DSSTTV scheme shown for System L is the subject of further study. All the schemes use 728 kbit/sec near instantaneous companding incorporated in sound coding, the same as that for the MAC Packet standards family for European DBS.

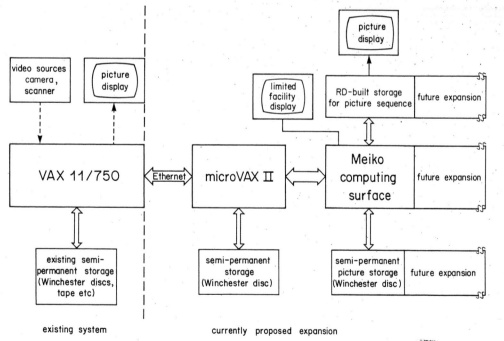

Fig. 5. Computer image signal processing experiments in the BBC are currently undertaken mainly on a VAX 11/750 served by comprehensive picture storage systems. A major expansion of the facilities has been proposed to exploit the substantially greater processing power available from the transputers in a Meiko computing surface to make HDTV signal processing research much more practicable. The proposed expansion includes picture sequence storage using RAMs to give full-picture display sequences of several seconds of 625 line color signals and, initially, 1 or 2 seconds of full-quality HDTV. The design of the development crucially includes provision for future expansion of transputer processing power and displayed sequences.

135

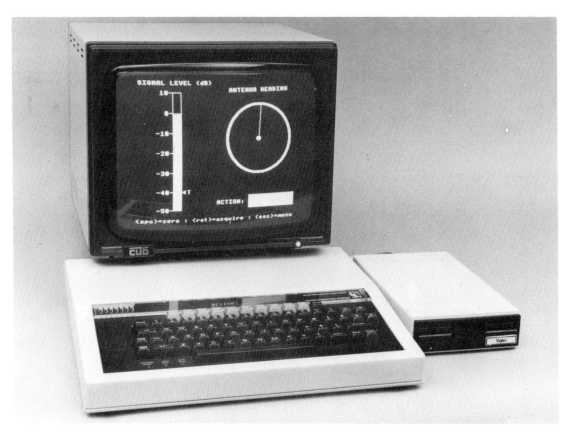

Fig. 6. Much of the BBC Research Dept.'s television work is concerned with current operational problems. An auto-tracking antenna system for a man-pack camera on a radio link has been devised, mainly for sports outside broadcasts. The photographs show field trials taking place at Wembley Stadium, London, together with the base station equipment which uses a BBC microcomputer.

Bruce Moffat graduated in engineering science from Oxford University in 1959. He remained at Oxford to carry out research in microwave electronics, for which he received a D.Phil. degree. He joined the BBC in 1962 and worked in the Acoustics Section of the Research Dept., where he was chiefly concerned with the application of computer techniques to studio acoustics. In 1966, he transferred to the Television Section, Research Dept. and investigated the problem of head-clogging in videotape recorders and possibilities for digital video recording. After two years as R & D Manager with Ilford Zonal, a magnetic recording media manufacturer, he returned to the BBC in 1970, in the Baseband Systems Section in the Electronics Group, Research Dept., and was engaged in the development of the PCM system now used for the distribution of high-quality sound signals; he became section head in 1971. In 1976 he was appointed head of storage and recording section in Studio Group, and was appointed head of studio group in 1981. In 1984 he became Head of Research Department. He has served on several CCIR, CCITT, and EBU Committees, and is currently Chairman of CCIR United Kingdom Working Group 11H concerned with High Definition Television.

The Big Screen is Closer Than You Think

John R. Forrest
Independent Broadcasting Authority
Winchester, United Kingdom

ABSTRACT

Plans for domestic satellite broadcasting (DBS) in the 12 GHz band in Europe are well advanced. Services using the MAC-packet transmission standard adopted in Europe will commence this year. Over the next three years, a number of national satellites will be launched and there is the likelihood of additional pan-European services. Though it had been thought that implementation of high definition television was some way in the future, this highly competitive programme environment will spur the introduction of a higher definition television service suitable for large, wide aspect-ratio screens with stereo sound. An important feature of this evolution is the capability of the MAC-packet system to deliver this new form of television, while providing conventional pictures to existing receivers.

INTRODUCTION

Since the introduction of colour, relatively little has happened to change the format of television receivers. Screen size has increased, but is approaching a practical limitation with cathode ray tubes owing to weight penalties associated with the vacuum envelope and its support structure. Though 40" diagonal dimension displays have been developed, they are somewhat impractical for domestic use and 30" represents the more practical upper limit. Flat panel displays seem unlikely within the next decade to replace cathode ray tubes except for smaller portable sets, but projection displays are showing most encouraging development. The most significant evolution in television in recent years, notably in the UK, has been the success of Teletext with some three million receivers (ie about 15% of homes) capable of displaying this service now in use. There is steady progress in the UK also towards the introduction of high quality digital stereo sound to terrestrial television services, but the best advantage of this is only likely to be realised if a larger visual image can be associated with the sound separation.

The next quantum leap in television, analogous to the change from black and white to colour, must therefore be the introduction of the big screen. It is most unlikely, however, that this would totally displace the smaller conventional television receiver. Current trends are to the ownership of more than one television per household. A likely future scenario, therefore, is a big screen television for family viewing and a conventional portable set for individual use in bedrooms, kitchens and smaller rooms.

The big screen carries with it the requirement for a higher definition in the displayed picture. The optimum, determined from many subjective studies, would appear to involve a

138

scanning format of some 1,000 lines, non-interlaced or with frame rates higher than 50 Hz. Though it is possible that such pictures, requiring some 50 MHz of video bandwidth, may be distributed by optical fibre cables to the home in the more distant future, there are no broadcast transmission channels at present available to handle these bandwidths. Terrestrial UHF television channel bandwidths pose severe limitations on what may be achieved, but DBS channels bring sufficient bandwidth increase to allow an acceptable higher definition signal to be transmitted, provided some signal processing techniques for moderate amounts of bandwidth compression are invoked.

The significant key requirements for a big screen, higher definition television service are now in place:

- transmission channels in the DBS band which allow adequate, though not ideal, bandwidth;

- the standardisation in Europe on the MAC-packet family transmission standard which allows wide aspect-ratio higher-definition pictures to be produced by appropriate receivers;

- the commercial availability of compact domestic projection displays, capable of producing pictures at least 1m wide in 5:3 or 16.9 format, but also compatible with 4:3 format transmissions;

- the ready availability of film-based high definition programme material;

- the very important practicality of maintaining the reception capability of existing conventional 625 - line receivers.

EUROPEAN DBS CHANNELS

The 12 GHz DBS band was planned at the World Administrative Radio Conference in 1977. The 27 MHz RF channels will handle video bandwidths up to approximately 12 MHz. The closer it is possible, therefore, to approach bandwidth compression ratios of about 4:1, the closer it will be possible to attain the ideal high definition format.

MAC-PACKET TRANSMISSION FORMATS

The EBU has specified a family of MAC systems [1] which are inter-related. The video format is the same for all members of the family (Fig. 1), but different formats are used for the sound and data parts of the signal. C-MAC is optimised for satellite transmission in the WARC '77 channels and uses a 20 Mbits/s data rate for sound and data. D-MAC is a similar data capacity format optimised for moderate bandwidth cable networks. D2-MAC is a format with half the data rate of C and D, namely 10 Mbit/s, developed for existing narrowband 7 or 8 MHz cable networks.

A considerable amount of work has gone into the detailed realisation of chip sets for D2-MAC receiver systems because of the cable application. Though this format does not allow the capabilities of the satellite channel for high definition television to be realised, the French and German DBS will use D2-MAC initially because of the short time-scales prior to satellite launches and commencement of service. With the longer interval before the launch of UK DBS in 1990, there is adequate time to bring into production receivers which can operate with the higher data rate of C or D format, thereby permitting initiation of an HDTV service. Fig 2 shows the parts of the signal format that are available to be used for enhancing the standard definition picture.

SIGNAL PROCESSING TECHNIQUES

A variety of signal processing techniques to achieve the necessary bandwidth reduction are being investigated, many of these being co-ordinated through EBU Working Groups.

(a) Scan Conversion

In many cases it is possible to reduce the line rate of a television signal without proportionately reducing the picture quality. Figure 3 shows the principle. A digital pre-filter and post-filter provide the interfaces between the reduced transmission scanning standard and the higher scanning standards at source and display. This type of approach has been central to the enhanced C-MAC studies [2]. So far in this work,largely due to the ready availability of 625 line displays effort has been concentrated on conventional 625/50/2:1 signals converted to a progressive 625/50/1:1 signal to give an enhanced picture. The techniques, however, are specifically relevant to the reduction of 1125 or 1250 line scanning format at source to 625 lines for transmission, followed by display at 1125 or 1250 lines.

(b) Sub-Nyquist Sampling

Sampling the television image at sub-Nyquist rate clearly provides bandwidth compression, but leads to aliasing problems. Suitable filtering of the signal prior to sampling can eliminate this problem except that there can be problems with artefacts in moving images. The MUSE system is an example of this approach [3], but has in existing implementations been incompatible with existing receivers

(c) Digital Assistance

In digitally-assisted television, (DATV), the transmission channel is effectively split into two distinct parts - the analogue conventional signal and a digital signal which describes additional features of the analogue information [4]. The digital signal could represent, for example, the motion content or spectral nature of the scene. However, the computational complexity associated with these techniques can be very high and it is clear that considerable research needs to be carried out before they can be implemented in a practical system.

All these techniques are more fully described in a recent paper by Long and Stenger [5].

THE IMAGE SEQUENCE PROCESSOR

In order to be able to evaluate rapidly the relative merits of various complex methods of picture processing, an Image Sequence Processor has been developed by the IBA. This is, in essence, a means of simulating any proposed video processing technique. It avoids the need for the involved task of constructing dedicated hardware for each technique to be assessed.

This machine has been designed very much with flexibility in mind, as parameters of processing techniques under study tend to vary widely and are not necessarily related to parameters of current systems. The processor has a large semiconductor RAM image store currently being increased to 256 MBytes in 8 bit words and a powerful computing system including an array processor which can rapidly manipulate the image data. There is 12 bit A/D conversion at input and output with 16 bit internal processing. A block diagram is shown in Fig 4. The image memory has enough storage to hold several seconds of HDTV moving pictures. The system can also read and produce tapes to the COST 211 half-inch magnetic tape standard, allowing image exchange with other organisations working in this field.

Image Memory using RAM rather than fast magnetic disks was chosen for speed of processing and for reliability, as well as ease of interfacing with the computer system. The 68020 processor used can address up to 4 GBytes of RAM directly, allowing for memory extension if required.

Techniques under investigation can be broadly categorised under the headings of "spectrum folding" and "motion-related processing".

"Spectrum folding" techniques cover the various spatial and temporal downsampling and filtering techniques mentioned earlier, basically involving the trading of spatial and temporal resolution dynamically according to the statistics or characteristics of the picture. These techniques can often be improved in their subjective temporal resolution by sending some extra digital information about motion occurring in the picture. This part of the study is covered under the "motion related processing" category. The study of motion-related processing including movement estimation is a very large subject and involves very complex algorithms, for which the theory overlaps that of artificial intelligence and requires a powerful computer to simulate. The need to compare the incoming image according to a complex criterion with many spatially shifted versions of the same image is a relatively easy task for the human brain, but by no means easy for a conventional digital computer!

The Image Sequence Processor can be used to simulate the whole processing chain from camera output to display input. A sequence from an HDTV camera operating in, for example, 1250/50 format, can be captured in the image memory. The computer can then spatially and temporally filter and process this signal, resulting in a near conventional 625/50 format compatible with normal receivers. This may be viewed to check for filtering artefacts, the inverse process applied, and any motion-related data incorporated before the picture is again viewed, in 1250/50 format, to evaluate overall transmission quality. Other line and field rates can easily be investigated as these are fully programmable on the processor.

In parallel with the immediate work for HDTV system realisation, a longer term investigation into bit-rate reduction for digital distribution is in progress on the Image Sequence Processor. Methods being investigated include transform coding and rely on coding the picture in such a way as to alter the statistical properties of the picture information. If the algorithm is chosen carefully, then various coefficients in the algorithm are zero or near zero and can be discarded for transmission, giving bit-rate reduction without unacceptable impairment to picture quality.

The flexibility of the Image Sequence Processor allows many other investigations, for example, standards conversion, RGB-YUV matrixing optimisation, gamma correction, and investigation of the subjective effects of loss of resolution in the spatial, temporal and chromatic fields. In particular, investigations into alternative matrices to YUV, colorimetry, and gamma-related phenomena may result in new methods giving useful reductions in data or bandwidth which are an advantage in transmission. The Image Sequence Processor is capable of storing un-gamma-corrected pictures as it has modes which allow twelve bits per pixel at input and output and sixteen bits per pixel internally for processing.

Source pictures for all these investigations can be captured both from facilities at Crawley Court, from HDTV equipment and from other organisations via the COST 211 digital tape standard. This latter aspect is particularly important to the co-ordination of this work with other European organisations through the Eureka HDTV project, the aim of which is to develop in a unified manner all the component parts of an HDTV chain, which would be compatible with existing 625 line systems.

CONCLUSIONS

The realisation of an HDTV service in the near future using the presently allocated 12 GHz DBS band is now feasible using the MAC-packet transmission standard. It does, however, require the use of a bandwidth compression technique and for this there are several contenders. The IBA image sequence processor allows rapid comparison of the available techniques and is a vital component in the programme to achieve, within the next two or three years, a fully developed HDTV system compatible with existing receivers.

References

[1] EBU Specification of the Systems of the MAC/Packet Family. EBU Doc. Tech. 3258 (October 1986).

[2] G. Tonge. "Signal Processing for Higher-Definition Television". IBA Technical Review No. 21 (1983).

[3] Y. Ninomiya et al. "A Single Channel HDTV Broadcast System - The MUSE". NHK Laboratories' Note No. 304 (1984).

[4] R. Storey. "HDTV Motion Adaptive Bandwidth Reduction". Proc. IBC (1986).

[5] T.J. Long and L. Stenger. "The Broadcasting of HDTV Programmes". EBU Review No. 219 pp 297 - 314 (October 1986).

Acknowledgement

The Image Sequence Processor is a major project in the Video & Colour Section of the IBA's Experimental & Development Department. Mr. T. J. Long, Dr. M. D. Windram, Mr. R. Morcom, and Dr G Tonge kindly provided background information for this paper.

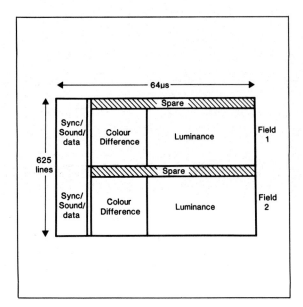

Fig. 1. Format of MAC signal.

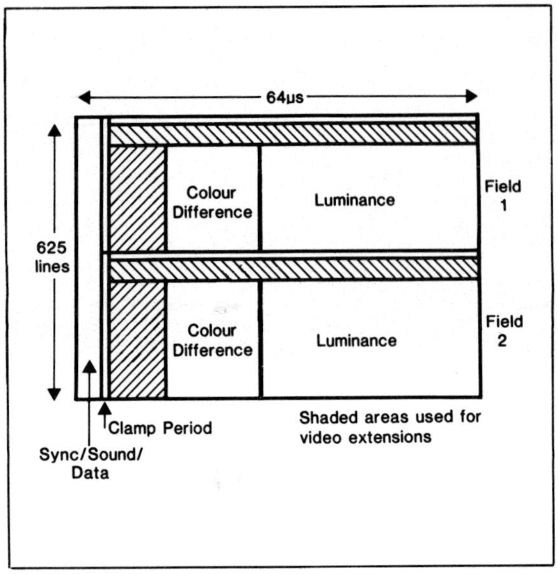

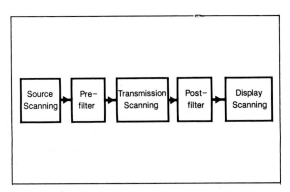

Fig. 2. Format of MAC signal showing parts available for provision of the HDTV enhancement.

Fig. 3. Bandwidth reduction for transmission in HDTV chain.

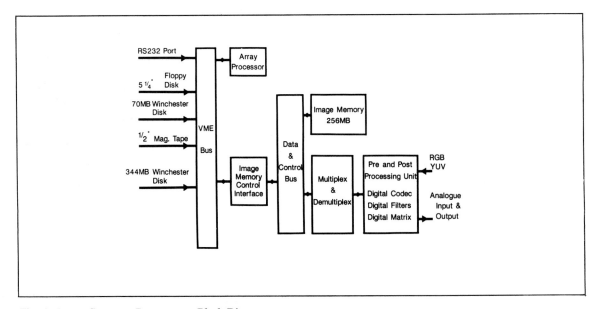

Fig. 4. Image Sequence Processor — Block Diagram.

John R. Forrest obtained his B.A. degree in Electrical Sciences in 1964 at Cambridge University. His postgraduate work for the D.Phil. degree was at Oxford University on the topic of Plasma Diagnostics for Nuclear Fusion. Between 1967 and 1970 he was at Stanford University, Calif., as a research assistant and lecturer. In 1970 he returned to take up a post at University College, London, becoming professor of electronic engineering in 1982. His research group concentrated its activities on new techniques in radar and satellite communication systems, particularly involving microwaves and optics. In 1984 he became technical director of Marconi Defense Systems, but left in 1986 to join the broadcasting world as director of engineering at the Independent Broadcasting Authority. He is Deputy Chairman of the IEE Electronics Div. and a senior member of the IEEE. In 1985 he was elected to the Fellowship of Engineering.

TV Research in CCETT: To and Through the 90s

Jacques Sabatier and Jacques Poncin
CCETT
France

1 - INTRODUCTION

This paper intends to give a view on the major tracks presently considered in France on the way to the future of broadcasting systems. The main research laboratory active in this field, CCETT, will first be presented. Then, some general considerations will be introduced on the way the broadcasting system is evolving, on the present situation in Europe and on the directions which, in our view, have to be explored prioritarily. Some of the research programs undertaken at CCETT on these main axes will finally be outlined.

2 - WHAT ARE CCETT LABORATORIES

The name of CCETT means in French : Joint Research Laboratory on Broadcasting and Telecommunications. A rather specific and unique feature of this laboratory is that it is a joint venture between two public authorities who play rather complementary parts in the field of communication in France : One is the PTT Administration which operates, up to now in a monopolistic situation, all telecommunication networks and services throughout the country, except for radio and television. The other one, Telediffusion de France (TDF), is a public company, which is in charge of broadcasting radio and television signals produced by public or private program companies, by the use of terrestrial transmission networks and transmitters, and satellites. By the way, the general and international deregulation trend will certainly introduce in the short future rather important modifications in the status and in the working environment of both companies, leading to a more and more competitive situation, but not in their general assignements, so that their motivation to jointly carry some research activity will probably last, as during the past 15 years.

All working of CCETT is based on a simple 50/50 principle : Each parent provides half of the staff and budget and holds equal power for orienting the studies. Both own equal property rights on any kind of results produced by the laboratories.
Presently the staff is stabilized at the level of 400 people and the laboratories are equiped with modern research facilities, including a powerful computer center, studio equipments for picture and sound origination, CAD and simulation systems, access links to national and international radio and satellite networks, etc ...

The work of CCETT is devoted to the development of all kind of new audiovisual communication services which may generate traffic on the networks operated by its parent organisations and offer them commercial diversification opportunities.

In a very broad view, such a development relies on the design and integration of several technical systems which constitute the program and signal chain, between the producers, or information providers, and the end users in the general public.

On a functional basis, 4 main sectors may be identified in such a chain :

- the production area including signal origination and editing,
- the head end storage and access system including data (picture, sounds, ...) bases and information retrieval sub-system,
- the network section providing the information transmission path between the head-end and the users,
- the terminals area including the end presentation units as well as program selection or interaction devices.

It is worth noting that this functional decomposition applies to broadcasted as well as interactive services and that the structure is rather independent of the nature, bandwidth, or bit rate of the information support signal.

In addition, transversal aspects have to be considered, like information representation (coding and transcoding along the chain) or communication protocols (for information transfers between the functional areas, or for additional functionalities implementation, like access control capabilities).

The analysis provides a view on CCETT laboratories organization, where roughly 25% of the research ressource is devoted to terminals, 25% to distribution networks (radio and cable), 10% to production tools and 40% to head ends and transversal aspects.

Although the basic technical research organisation is not service but function-oriented (this allowing a cross fertilization and coherence between the studies of various services), a strong coordination is maintained in order to enable the building up of complete technical validation chains for each service. And this technical assembly job, realised in project operations, is often extended to the setting up of field trials, where operational aspects may be evaluated, where a first glimpse of service usage may be caught, and where contacts with potential users (especially on the program providers side) may be initiated.

All this work implies of course a lot of cooperation with external partners. In the first place, industry has to be associated at a very early stage to the developments of various parts of the systems, and a lot of discussions with other bodies are necessary, in the framework of international harmonization or standardization organisations (CCIR, CCITT, ISO, EBU ...), for the setting up of standards. Contribution to standardization work has been for a long time an important part of activity of CCETT, but cooperation at the level of research work itself increased significantly during the last 2 or 3 years, with the launching of european cooperative research programs like ESPRIT, RACE and EUREKA.

Concerning the precise nature of activity of CCETT staff, it is worth focusing on the importance of technical support activity, which, on one side, directly benefits to its parent organisations and, in return, provides to CCETT prospective studies a strong anchoring in the concrete environment : it first consists in assistance for specifying and supervising the industrial developments of all key elements which must be available, both on the professional and on the consumer side, for a new service to be opened ; it also implies various forms of know-how transfers to operational structures, especially through extensive field trials. This has been done, for example, in the recent past, in the field of D2-MAC/Packet system development for direct satellite television broadcasting or for the telematics or interactive videotex system development which has come in France to an unequalled size and success.

As a conclusion, it is clear that CCETT research activity is not oriented to fundamentals aspects or basic technology (althought efficient use of up-to-date technology implies some test or evaluation activity), but to short and medium term system development.

3 - TV SERVICES EVOLUTION

Communication services, and among them TV services, have evolued since 40 years according to common economical laws. This very straigh-forward fact is sometimes masked because engineers have a natural tendancy to look at these questions from a technical point of view, and to forget that TV is first of all a business.

3.1 - Evolution in the past

The weight of distribution networks and of user equipment is such that it limitates considerably the way in which communication services can develop. A change in distribution systems can be thought of, when there is a good chance that the new form of audiovisual product will be accepted (and paid for) by the public. A deep change is considered only when there is an actual need for a modification because the system previously used cannot cope with new requirements. When compelled to change a distribution system any operator will first provide services which proved to be profitable and then think of offering another type of service. It would be extremely risky for him to decide to change at the same time the distribution system, which calls for heavy investments, and the nature of the service, without any idea of the success it may get from the public. It is an obvious statement to note that the economic viability of a service should first be likely and that any "plus" in the service, which requires extra equipment can only be introduced if the quality/price ratio is acceptable by the end user. Three examples can be given.

- When the first monochrome TV systems were designed, it was necessary to adapt the standards to the capabilities of the existing technology and to the low bandwidth implementable equipments (interlace scanning has been introduced at that time). These decisions permitted to put in operation monochrome TV networks and to sell TV sets to the first spectators.

When colour TV appeared, there was a choice between a completely new system (sequential colour system) unreceivable on monochrome sets and a compatible colour system (NTSC). The latter solution has permitted to introduce smoothly colour programs without loosing the existing audience. The distribution networks were kept and rensed while the characteristics of the service were changing. The same route has been followed some years later by European broadcasters.

- Cable networks developped to distribute programs in areas where they could not be received in good conditions and to bring to the homes more programs. When the first investment has been made then the evolution of services has to preserve this heavy first expense and to use it in the best conditions : by increasing the number of programs, new selectors are needed by introducing some interactivity for answering live questions or voting in real time (a special return channel is implemented) ; by developping pay TV services based on a scrambling process and a conditionnal access system (specific decoders are provided)... All these improvements are made in normal commercial conditions. As time goes it may happen that the maximum usage of the capabilities of the first network cause the complexity and volume of the equipment in the households to become very important ; the operator may then to think of changing the configuration of the network to try to make an overall economy by centralising or even eliminating a number of equipments. For instance a star distribution system can provide interactive and especially pay TV services in a much more flexible way than a tree system. Some network operators are faced with such a problem. Of course the opportunity of this change would also be used to design a network able to be a new start point for a new evolution in the nature of services.

- Deep modifications in a distribution system can nevertheless happen in very particular conditions. When the compact disk has been introduced, it was commercially presented as a revolution in the HIFI business. A new equipment was needed to read these new disks, in connexion with existing HIFI amplifiers and loudspeakers. But there was a public educated in music listening, who had moved from AM to FM, from traditionnal to HIFI equipment. The product was successful because it was bringing to the user a program he knew (the same opera, concerto, symphonia at a price considered as fair given the remarquable increase in quality).

3.2 - Conditions of the technical evolution

This very progressive evolution shows a strong interference between signal specification, bearer characteristics and service objectives. The interaction between these three items is the engine of the evolution of communication services. An example can be given, concerning present research activity. The idea of changing from composite TV systems to MAC packet systems in Europe resulted from the conjunction of at least three factors :

- DBS is basically a transfrontier service which requires a unique coding scheme of the signal,
- FM broadcasting decided in DBS because of the specific properties of the channel increases the quality of the received signal and gives the opportunity to reach much better performance than with PAL/SECAM even when used in the best conditions,

- Digital coding techniques for sound allow new facilities like HIFI sound, stereo, multisound and multilangage.

The discussions resulted in the C/D/D2-MAC-Packet systems, each of them having different capabilities. The choice of one of these systems is left to the operator and strongly depends on the bearer available to him at the time he wants to exploit the system.

So the availability of a new distribution system can induce an evolution in signal specifications.

Vice versa a new signal specification can induce evolutions in services. For instance, D2-MAC-Packet can be used on terrestrial broadcast networks and on cable networks with a frequency spacing as low as 7 MHz. It is not possible to receive it directly with a normal TV set. But, because pay TV services use a special decoder, they can move to D2-MAC Packet which has efficient scrambling capabilities. Besides a considerably improved resistance to piracy, the consumer would benefit of a better picture and a better sound quality.

DBS projects have also generated a lot of work on distribution techniques. For example distribution of DBS programs is possible in MATV with 2000 MHz distribution cable networks using the first intermediate frequency chosen in DBS equipments (950-1750 MHz). This solution is much cheaper than foreseen and offers 40 extra channels in FM.

It's clear that, without DBS, this possibility would have perhaps never been explored ; it allows to think of new service possibilities, because there will not be 40 DBS programs in the short term.

When comparing systems and making evolution forecasts it is of paramount importance to replace correctly the events in time.

- at the level of networks : when will the 12 GHz DBS systems be in operation ? The 22 GHz DBS systems ? The high data rate digital distribution cable networks ?
- at the level of recorded media : when will high capacity disks or tapes be available at a consumer price ?
- at the level of display technology : when will a one square meter HDTV display be available at a consumer price ?

The interaction happens between service-signal-bearer, when they are coincident in time. The order in which they are expected to be available has a considerable impetus on the nature of the evolution and on the standards themselves (see example in paragraph 4.1).

3.3 - An evolution model

A theoretical model of this evolution can be proposed at this stage. In a three axis system, each refering to one of the three basic characteristics previously described (signal representation, bearer characteristics, service definition). A commercial situation is represented by a point. When an operator has the project to move from one situation to another, the move on the service axis is a measure of the expected interest of the consumer and, the move on the signal axis and on the distribution axis is a measure of the price to pay in equipment.

The common economic limitations will apply : the quality/price ratio must be acceptable by the public, and the overall expense supported by the operator must be at least compensated by the expected revenue. So when a project of a new service using special signal coding and distribution characteristics is such that it is too "far" from any existing profitable situation it is necessary to modify the description of the service to reduce its "distance" to a known situation and perhaps prepare a path across several commercially profitable situations to reach the target of the project. If there is no continuous path possible, the project is likely to be economically non viable.

In the evolution of TV services the start point is very important. From one country to another, it's easy to note a very different degree of development of TV services, that is different start points for new evolution. TV operators who would wish to reach a similar situation in different countries have to find viable routes in economic terms to move from the actual situations to the target situation. Such routes don't necessarily exist for all of them and they may have to change the definition of their target. This remark shows, that may be a conflict between the understandable wish to unify as far as possible the technical charateristics of systems and services all over the world and the necessity to ensure viable evolutions in each country.

3.4 - TV services in Europe

In most countries in Europe, the viewer has a limited choice of programs because Television has developped as a public service controlled by governments with the aim of a 100 % coverage of the country. This planning method uses a large amount of frequencies and it is quite difficult to find room for new programs even on local basis in the frequency band. In some countries this statement no more applies : in Italy a large amount of private TV stations have developped without constraints, each chosing its own broadcast frequency ; in Belgium and Netherlands which are smaller countries a large number of households are connected to cable netwoorks which retransmit programs from neighbouring countries. In spite of some move in other countries, this picture is likely to be valid for some years.

3.4.1. - The interest in DBS

DBS is a way to increase significantly the number of available programs received in areas where the density of population is too low or the size of villages too small to justify a cable netwoork. It's a reason why DBS is supposed to be successful in same European countries.

Dedicated antenna and Tuner are needed to receive DBS signals. It is generally considered that the power of DBS satellites like TDF1 or TVSAT is such that the price of this equipment should not be a blocking factor.

PAL with an FM subcarrier for sound. Work made till now in the EBU has led to the specification of the so called MAC-Packet Family of standards, among which the D2-MAC-Packet member has been chosen by the French and German Governments. (In Europe there is no more choice between PAL/SECAM or MAC-Packet since the Council of Ministers of EEC has decided the latter to be the choice to make).

As far as research is concerned, basic DBS is no more a subject of study. The DBS system is taken as a startpoint for new evolution.

3.4.2 - Evolution towards HDTV

How to distribute HDTV to the public is then the main question for broadcasters, who know that they will have to compete with other media like cables, tapes and disks. A large amount of views have been expressed, and there is a hard debate between revolutionnary and evolutionary approach defenders :

- the evolutionnary approach is based on the concept of operational compatibility at the level of receivers. It means that an HDTV broadcast system is designed on top of a basic D2-MAC/Packet signal, in a WARC 77 Channel ; the picture is displayed with the MAC quality by a receiver designed to receive the basic D2-MAC/Packet signal without modification, and with HDTV quality by a receiver specially designed for that.

- the revolutionnary approach aims at defining a new signal optimized independantly of any compatibility constraint other than channel constraints ; dedicated receivers are needed to display the HDTV program.

The "revolutionnary" approach has to face a number of difficulties :

- the price the user would have to pay to buy a completely new equipment, is likely to be very high ; the audience would probably grow very slowly.
- the operator should provide programs in HDTV only which needs a reasonable stock of programs and important means for live programs production.
- a HDTV display adapted to domestic usage should be available.

The evolutionnary approach doesn't suffer from the first two drawbacks because of its inherent progressivity. It is clear today that there is an enormous difference between the wiewing experience one can have in front of PAL/SECAM TV picture and in front of a studio HDTV picture on a large screen at 3 H viewing distance. But it is necessary to recall that comparisons must be done at the time when the viewer will make his choice. The consumer of the end of the century will have to compare two performances :

- The optimum HDTV system (with the low but unavoidable degradation due to the coding process),
- The DBS picture in which facilities like 16/9 aspect ratio or post processing at display level will have been introduced (this performance is of course far better than the one taken as a reference today).

An extra processing will provide an HDTV quality, (perhaps slightly lower than in the first case). The gap in quality is likely to be much too small compared to the price of a complete new equipment and a reduction in the access to previously broadcast programs.

The revolutionnary approach is more and more considered as unrealistic. But it doesn't mean that all engineers have in mind the same kind of evolutionnary scheme. That is going to be the debate of the years to come.

4 - MAIN AXES OF WORK AT CCETT

CCETT is presently involved in research programs along the three main axis used to represent a service situation in paragraph 3.2. In each domain some flexibility is maintained so that the research teams may explore various fields where innovation may spring, but most efforts are devoted to the areas where economic viability for the systems is foreseeable for the medium term.

4.1 - Evolution of signal quality and representation standards

CCETT is involved in TV signals analog as well as digital coding development. The description of the MAC signal is now completed in the DBS standard. The work is thus now focused on the move to HDTV distribution standards, considering two complentary approaches : on one side, sub-sampling in order to make optimum use of the WARC 77 analog FM channel and on the other side the DATV principle for carrying to the receiver the extra information needed to built up an HDTV picture. The progress made during the last year in both domains is such that there is now a good chance that the feasability may be demonstrated and a prototype built in the short future. Howeverr, the editing of the detailed standards, which is mandatory before distributing signals to the public, and the design and manufactoring of receivers to cope with this evolutionnary situation, will probably be finalized only in the first years of the 90's.

Important effort is also devoted to bit rate reduction techniques applied to the 4-2-2 signal. The first objective is clearly the design of an exchange system fitting with the third level of the European hierarchy, i.e. 34 Mbit/s and offering some downstream processing capability. A number of international study groups are working on this subject and plenty of coding schemes are explored : all kinds of DPCM, transform coding and mainly DCT. Lower bit rates in the range of 10-20 Mbit/s are also considered as relevant objectives for distribution where the design constraints (e.g. post processing) may be alleviated and where picture quality is the only criterion to be taken into account.

Redundancy reduction of HDTV signals is also an important step to be jumped before the mid 90's. Although it is still difficult today to forecast the final results, two ways are offered for reducing the bit rate to 100-140 Mbit/s :

- one is based on digitizing the MAC signal (PCM coding) and adding the low bit rate DATV signal. (Bit rate reduction may perhaps be applied to the PCM MAC coded signal but one should not forget that DATV processing already made use of a redundancy which is no longer available.

- the other method would be based on straight coding of a digital HDTV signal, with techniques extrapolated from those used to carry normal TV with 10 to 20 Mbit/s.

If the evolutionnary approach to HDTV in DBS succeed, the first solution would be the most attractive as it is a simple translation of the analog signal in digital form, allowing to rense all the terminal equipments. If it fails, the second solution is an alternative. Once again, it may be noted that the technical decisions to be taken by the end of the 90's, will strongly depend on the routes which will have been followed previously.

4.2 - Evolution of distribution technologies

TV broadcasting started developing in the VHF and UHF bands and is now going up to 12 GHz DBS. The WARC 77 frequency plan will be used to transmit the MAC-packet signal. When Carefully studying co-channel and adjacent channel interference, and cross polarization protection, and taking into account possible reception technology improvement, it appears that a significant increase of baseband signal bandwidth is still possible. This is probably a key towards compatible HDTV broadcasting.

Another another path to HDTV-broadcasting is 22 GHz band. This band is not yet affected to DBS in Europe. But studies are carried on by all broadcasters to prepare the 1992 conference. The main difficulty in preparing such a planification is the choice of a good model of HDTV signals, considering that the operation of 22 GHz systems is unlikely to start before the end of the 90's.

The recent evolution of TV distribution techniques, the DBS projects and the emergence of low bit rate integrated service digital networks (ISDN using 144 Kbit/s connection link) in the telecommunication area, result in a general reflection on the best way of cabling the homes. In the future : one may question whether a kind of multiservice integrated distribution network should be installed or whether different networks should still be laid down side by side waiting clearer needs to integrate some parts. Once again, economy will decide. Our present studies encompass coaxial and optical fiber technologies, many kinds of network architectures (tree, star, loop, bus,...), signal specifications, interconnexion protocols and... legal considerations.

We have no specific technical studies on recorded media, like disks and tapes. But we keep informed of the evolution of these techniques to evaluate their capabilities : in commercialization many cases indeed, there may be direct concurrence for audiovisual products between recorded media and network distribution and it is obviously vital for PTT and TDF to correctly evaluate such a competition.

4.3 - Evolution of audiovisual services

Apart form the trend to continuously increase the quality of the signals broadcasted to the public, a major evolution modern technology will induce in audiovisual products distribution, will be in the field of program access conditions : let us develop two examples :

- The first one relates to elaborated access control techniques. Such techniques have been introduced for the "pay per view" service on the cable networks which is, at present, far more developped in the USA than in France. However, we experienced, in the recent years, in our country, a rather unique system with successfull launching of a national radio-broadcasted pay channel using encryption techniques, which has now 1.5 million subscribers and which will evoluate in the short future towards "pay per view" mode of working. At the same time, we are studying a rapid introduction of access control as part for the operational DBS service : the security, or non-piratibility of the system will be very much improved, compared with existing systems, through the possibilities offered by the MAC signal representation, especially because it is very easy to process digitally for the descrambling processing, without any signal degradation. This fonctionnality may be simply and cheaply added to first generation receivers.

Simultaneaously the various way for managing access control entitlements, distributing access rights, etc... are sutdied and, for the medium term, we give considerable attention to the possibilities of "smart cards" as protected information support media. It is indeed very probable that an important development of these cards, for broad public use, will happen, in conjunction with other application development like banking or transactions through telecommunication networks (telematics).

- The second service diversification way we are considering for the future, precisely relates to the coupling of TV broadcasting with telecommunications and especially with telematics. Telematics provides a very cheap and easy to use return channel and interaction terminal which are potentially available in every french home at the end of this decade. Many applications will certainly appear, rather spontaneously in the short future, but we already started, on an experimental basis, investigations on two services : the use of the return channel for allowing the viewers - to participate to the programming of channels on cable networks - to express opinions, in real time, during the broadcasting of any program.

Both services will probably flourish and diversify with the grow of cable networks in France.

CONCLUSION

TV research in CCETT takes a very large account of the conditions in which TV services can be introduced in the public. The basic studies are organised to keep up with the greatest number of technical events in the world but the main effort is put on subjects which have the greatest chances to lead up viable systems. As an example, there are some studies on high bit rate distribution of TV signals to the home, some ideas on digital coding of HDTV and some activity on the architecture of distribution networks ; but the general study of the distribution of digitally encoded HDTV signals on a star network is not a main project ; it is not clear today when and why these networks would be built and when digital HDTV technology is likely to be available.

A research center is a place where new systems and new technology are elaborated. Among the many ideas which are put forward in conferences and study groups, it is necessary to make choices and try to keep a good balance between what is selected to be done at home, what needs to be followed and what is decided to be neglected. The long development made in paragraph 3 can be surprising in a research center like CCETT but careful examination of how TV has evolved is a way of selecting research axis. This evolution of TV services is in fact made of steps and progressivity, sometimes small steps, sometimes much greater steps under permanent control of economic viability.

Jacques Sabatier was born in 1946 and is a graduate of Ecole Polytechnique (1966) and Ecole Nationale des Télécommunications (1971). In 1971, he joined the research department of the Office de Radiodiffusion Télévision Française (ORTF). In 1972, he moved to the Centre Commun d'Etudes de Télévision et de Télécommunications (CCETT) in Rennes where he managed several research activities on picture coding and transmission, image processing, and picture quality evaluations. From 1980 to 1983, as an assistant manager of Télédiffusion de France Laboratories in CCETT, he contributed to the satellite broadcasting research program, to the smart card development, and to cable networks research activity. In October 1983 he was appointed director of the CCETT. He is presently chairman of the EBU Sub-Group V1 dealing with television coding and HDTV systems. He is a member of the French delegation to CCIR Study Group 11.

Jacques Poncin graduated from Ecole Polytechinque and Ecole Nationale Superieure des Telecommunications. He started his professional life in the late 60s in the Centre National d'Etudes des Telecommunications (CNET). He worked on several projects in sound, speech, and picture coding. In 1972 he moved to the Centre Commun d'Etudes de Tele diffusion et Telecommunications (CCETT), where he has been in charge of a research department on digital television coding for transmission and production purposes. At that time he was a member of EBU working parties and took part in CCIR standardization study groups. In 1980 he was appointed head of a division covering several research departments such as subjective evaluation of pictures, picture coding, CATV services, packet switchers and local area networks. In 1983 he was appointed assistant director of the CCETT.

Improved Television Systems: NTSC and Beyond

William F. Schreiber
Massachusetts Institute of Technology
Cambridge, Massachusetts

ABSTRACT

After a discussion of the limits to received image quality in NTSC and a review of various proposals for improvement, it is concluded that the current system is capable of significant increase in spatial and temporal resolution, and that most of these improvements can be made in a compatible manner. Newly designed systems, for the sake of maximum utilization of channel capacity, should use many of the techniques proposed for improving NTSC, such as high-rate cameras and displays, but should use the component, rather than composite, technique for color multiplexing. A preference is expressed for noncompatible new systems, both for increased design flexibility and on the basis of likely consumer behavior. Some sample systems are described that achieve very high quality in the present channels, full "HDTV" at the CCIR rate of 216 Mb/s, or "better-than-35-mm" at about 500 Mb/s. Possibilities for even higher efficiency using motion compensation are described.

1. Introduction

The existing television broadcasting systems - NTSC, PAL, and SECAM - have spawned large and profitable industries; indeed they may fairly be said to have had a profound influence on modern society. Nevertheless, they are far from perfect. Most of the knowledge exists to improve picture and sound quality so as to equal or exceed that of 35-mm motion pictures. If we want it enough, we can have higher resolution, better motion rendition, a wider field of view, world-wide compatibility, and freedom from most defects.

Many, if not most, of the issues involved in improving existing systems or introducing entirely new ones, are primarily political or economic, and not technological. The main social value of a television system, economics aside, certainly resides in the programming and not the image quality. Interesting as these issues are, they lie outside the subject of this paper. We are concerned here exclusively with means by which sound and picture quality can be improved, and, to a lesser extent, with the means by which such improvements can be brought to the marketplace.

The psychophysical phenomena and signal-processing ideas underlying much of the following discussion were dealt with in an earlier paper and will only be summarized here. [1]

1.1 Why Now?

As equipment has improved and professional viewers have become more critical, the shortcomings of NTSC[1] have become more obvious, at least to TV practitioners. What they accepted 30 years ago, in the full flush of the successful design of a color system that did not obsolete existing receivers, is no longer acceptable. There certainly is no grass-roots demand for improvements, and there is little evidence that these perceived shortcomings are significant to the commercial health of TV. Nevertheless, there may well be commercial opportunities in the provision of significantly improved television services.

Rapid progress in semiconductor development, driven primarily by demands of the computer industry, has led to more powerful chips and lower prices. This presents the possibility of much more "intelligent" receivers and more efficient use of channel capacity. The existence of such capability, and its application in related areas, creates a pressure to make applications to TV, as exemplified by digital receivers.

Another factor creating a favorable climate for change in TV is the improved understanding of TV signal processing. Such processing has already been applied to graphic arts, generally at low data rates, but with excellent results. Demonstrations have been made showing good motion rendition with very few frames per second, [2] elimination of interline flicker by up-conversion, [3] and improved separation of luminance and chrominance by means of comb filters. [4]

No doubt the most important element in creating interest in this subject was the demonstration of the Japanese high-definition television system in 1981, after a development that took more than ten

---

[1]Most of what is said here about NTSC applies to PAL and SECAM as well.

years. [5] Orchestrated by NHK, with contributions from many Japanese companies, images have been produced that are comparable to 35-mm theater quality. The noncompatibility of that system, and the very wide required bandwidth, have sparked developments in a number of other countries, mostly directed to improvements that would be easier to implement. [6]

1.2 Limits to Quality in Present-Day Television

The image quality actually observed on home receivers is affected by every physical element of the system, as well as by the system design. Since it is spectrum space that is the limited resource, an argument can be made to ensure that *only* the bandwidth limit the quality. As a practical matter, this is not entirely possible. However, the extent to which other, correctable, factors degrade the quality at present is much larger than seems reasonable.

1.2.1 Interlace and Line Structure

The choice of lines/frame and frames/sec for a given bandwidth trades off spatial and temporal resolution. All modern TV systems use 2:1 interlace in order to raise the large-area flicker rate that results from this trade-off. Alternatively, the use of interlace can be considered an attempt to raise the vertical resolution for a given field rate and number of lines/field. The effectiveness of this stratagem is, at best, limited. [7] Interline flicker at the frame rate is produced along with artifacts due to vertical motion. The former becomes disturbing according to the vertical resolution of the camera, which, typically, is barely half what it might be with 525 lines/frame. [8] Ironically, if the camera resolution were raised to the alias-free Nyquist rate of 262.5 cycles/picture height (cph), the interline flicker would be unbearable.[2] While the artifacts could be avoided by using progressive scanning, that would also permanently enshrine the lower vertical resolution. As discussed below, a solution lies in separating camera and display scan patterns from that of the channel.

1.2.2 Interference Between Luminance and Chrominance: Resolution Compromises

The ingenious band-sharing principle [9] that made NTSC backward compatible does not, in fact, work perfectly with the kind of processing envisaged by the advocates of that system in the 1950's.[3] This is somewhat puzzling, since comb filters were well known for other purposes at that time [10], and, in fact, had been discussed by Gray in his 1929 patent, the earliest mention of frequency interleaving. [11] The pioneers seemed much more concerned with making the subcarrier less visible on the existing monochrome receivers (that had the full 4.2 MHz bandwidth called for by the existing standard) than on making color receivers work properly. [12]

Whatever the history, we are all now very familiar with cross color, the appearance of spurious color in regions of high luminance detail where luminance information is interpreted as color, and cross luminance, the moving cross-hatch pattern that develops at the boundaries of brightly colored objects where color information is interpreted as luminance. Cross color cannot be removed easily at the receiver, but cross luminance can be controlled by restricting the luminance bandwidth to well under 3.5 MHz, and often as low as 2.5 MHz. The latter procedure, of course, violates the principle of full backward compatibility.

Even without these defects, which, at least in principle, can be removed by suitable multidimensional filtering at both transmitter and receiver, the system design itself restricts horizontal color resolution to a very low value, typically about 1/7 that of luminance, and very much lower than the vertical color resolution. Although there is no need to have vertical and horizontal resolution identical, a 7:1 disparity is very visible. [13]

1.2.3 Performance of Cameras and Displays

Ideally, the response of cameras and display should be 100% throughout the Nyquist baseband. Because of some peculiarities of camera operation, the vertical response is much poorer than that. The temporal response is determined by camera integration. This blurs moving objects, but does not limit the bandwidth enough to avoid temporal aliasing. Use of a shutter to keep moving objects in focus makes the aliasing worse. What is needed is a camera with high and fully controllable resolution in all three dimensions, a result that can only be obtained by operating a camera of inherently high

---

[2]Another way of putting this is that if interline flicker is *not* observed, the camera has insufficient vertical resolution and is therefore wasteful of channel capacity!

[3]It also limits the alias-free bandwidth to $+/-7.5$ Hz, rather than $+/-15$ Hz, as in the monochrome system.

resolution at a line and frame rate well above that of the channel.

Displays have somewhat different problems from those of cameras. The line structure, which tends to mask fine detail, cannot readily be removed without reducing sharpness. The fact that this seems not be a serious problem in contemporary HDTV systems is further evidence that the camera and the display, and not primarily the system parameters, limit the vertical response. A corresponding effect is present along the time axis. Phosphor persistence short enough to avoid temporal blurring produces a large fluctuating component of brightness in addition to the (time-average) desired value. With normal frame rates, filtering in the human visual system sufficient to avoid flicker must necessarily reduce the response in the temporal Nyquist band.

These phenomena point up the fact that the camera and display not only limit the system resolution, they must therefore also limit the efficiency of utilization of the channel capacity. As long as this is the case, there is no point in providing channel capacity very much in excess of the capabilities of the terminal transducers.

1.2.4 Transmission Problems

The striking difference in image quality between TV receivers in Japanese and American hotels, both using NTSC, is clear indication that it is not just the system, or even the receivers, that limit image quality. Good pictures require good signals at the receiver terminals. Furthermore, it appears that the higher the resolution of the TV system, the more easily quality is lost by such factors as interference, channel noise, and multipath transmission. Especially in analog systems, there is no simple remedy to this problem, except to advocate a much higher standard of operation and maintenance of every part of the TV plant, from studio to the home. Should digital transmission ever really become the norm in delivery of signals to the home, the job would be immeasurably simplified. Perhaps that is one good reason to go digital. Important as this problem is, it is not the subject of this paper.

2. Improved NTSC

We firmly believe that the full capabilities of the NTSC system have not yet been exploited. While it might not be capable of "theater quality," if the defects are removed and certain other changes made in the studio and in the receiver, then pictures much better than we have ever seen should be produced.

There are many advantages to improving TV on the basis of the present systems. This evolutionary approach, in which existing receivers may continue to be used, albeit without all of the benefits of whatever improvements are made at the transmitter, ensures the presence of the audience, a necessary element for manufacturers, producers, and broadcasters alike to be willing to make the required investments. It also avoids the politically untenable situation in which viewers find their receivers no longer of use because the broadcasting standards have been changed. The disadvantage of this approach is that there is an upper limit to the quality that can be attained, and this limit may well be below what is desired. In addition, even if full theater quality could be achieved in a compatible manner, it is not clear that a sufficiently large portion of the audience would buy new and more expensive receivers just for the sake of higher technical quality. If this were the case, the economies of scale required to produce sophisticated receivers at acceptable prices might never be reached.

It is possible to envisage a scenario in which compatible improvements were introduced over a period of years, during which time the shape of a final, noncompatible system might emerge. By this time the audience would have become more knowledgeable about quality issues and more discerning in their choices. A line of multistandard receivers could then be developed, capable of receiving new broadcasts as well as NTSC, and a new broadcasting service with more attractive programs inaugurated. Old receivers would continue to be serviced by down-conversion to NTSC of at least some of the new broadcasts. This would provide an audience for the new programming as well as programs for the existing audience. At some point, announced several years in advance, the down-conversion would stop. If it were desired to keep providing some service for the older receivers, special broadcasts of a limited nature could be provided for an additional transition period, as was done in Britain when the shift was made from 405 lines to the CCIR standard.

2.1 Removal of Cross Effects[4]

Even though Fourier analysis dates from 1822 [14] and was ably used by Mertz and Gray [15] in their 1934 analysis of the spectrum of monochrome video signals, it was not properly applied to NTSC signals until 1967. [16] It was not therefore available to the NTSC in 1953, where it would have made much clearer what was involved in proper separation of chrominance and luminance. The full derivation of the spectrum of composite signals is given in [17] and outlined in the appendix. At this point we simply note that since the light intensity on the focal plane of the camera is a function of x, y, and t, its spectrum is three dimensional, the axes being labelled horizontal, vertical, and temporal frequencies. As shown in Fig. 1, the subcarrier corresponds to spatiotemporal harmonics located in four of the eight corners of 3-d frequency space, the chrominance sidebands being located around the subcarrier in the same way that the luminance sidebands are located around the origin.

To be separable, chrominance and luminance harmonics must occupy distinct portions of frequency space. Normally, the respective video signals are bandlimited horizontally only, and not vertically or temporally except by natural characteristics of the camera. It can readily be seen from Fig. 2 that, if both signals have full bandwidth in these two directions, overlap must be present and cannot be removed by any operation at the decoder. To avoid this irreversible mixing, filtering is required at the transmitter, before encoding.

The simplest filters to use are one-dimensional, in which case they may be implemented as ordinary RLC networks and applied to the video signals. If luminance is limited to 3 MHz and both chrominance components to .6 MHz, i.e., if we give up temporal band sharing, then cross effects are removed. This approach gives up about 30% of the luminance bandwidth and about 50% of the I-signal bandwidth.

Better performance can be achieved with 2-dimensional processing, normally implemented with transversal filters[5] using a number of line- and sample-delay elements. As shown in Fig. 3, these can be used to cut the corners out of the 2-d spectrum. Note that it is necessary to limit the vertical resolution of chrominance to perhaps half that of luminance in order to have reasonable vertical luminance resolution at high horizontal frequencies. Note also that "diagonal" luminance resolution[6] must be given up to make room for chrominance. This is generally a good deal less damaging to image quality that giving up horizontal or vertical resolution. Excellent results with such filters have been achieved at INRS-Telecommunications in Montreal. [18] Another approach, which is intended to permit the use of simpler 2-d filters, is to make them adaptive. [19]

With 3-d filters, which require frame delays for implementation, it is, in principle, possible to maintain the full NTSC luminance and chrominance spatial resolution at low temporal frequencies. As shown in Fig. 4, this is done by limiting the spatial resolution of both components at high temporal frequencies. This is generally harmless except in the case of visual tracking of moving objects, in which case high temporal frequency response is required in order to preserve their sharpness. So far, this procedure has never been optimized, so that it is not possible to say how well it may work.

One possibility, with either 2-d or 3-d filtering, is to utilize horizontal resolution of the I component higher than that of the NTSC standard by giving up some horizontal luminance resolution at high vertical (or temporal) frequencies. This would cause worse cross effects on existing receivers, which might, in turn, be ameliorated with some kind of adapter.

In designing a new composite system in which 2-d or 3-d filtering were the intended mode of operation, it might well be possible to use a lower subcarrier frequency, at about 3.2 MHz, to allow double-sideband quadrature modulation of two color components, each of 1 MHz bandwidth. If the vertical color resolution were then limited to half that of luminance, a more symmetrical arrangement would be achieved.

2.2 The "Fukinuki Hole"

---

[4]This material is taken from a forthcoming tutorial paper by Eric Dubois and the author.

[5]Note that all transversal filters are "comb" filters, i.e., they have a periodic temporal frequency response. The usefulness of comb filters in separating luminance and chrominance, which have interleaved periodic spectra, is obvious. However, efficient separation requires using a rather large number of line delays. The usual one- or two-line comb filter does not separate the signals very well, without at the same time substantially reducing vertical resolution, at least at the higher horizontal frequencies.

[6]The use of this term is associated with the zone plate test pattern. Such components also result from two-dimensional texture.

In the 3-d approach to the spectrum, it is evident that, since only four corners are occupied by chrominance, the other four can perhaps be used for another signal, leaving the luminance spectrum perfectly symmetrical. This requires a second subcarrier, f_h higher or lower than f_{sc}. Fukinuki [20] first proposed using this extra channel for higher luminance components, and has recently made a demonstration of the process. [21] It should be noted that for the "hole" to exist, the vertical chrominance resolution must be no more than about 131 cph, which is only *half* the vertical Nyquist bandwidth, as shown in Fig. 5. It may well be that the overall image quality would be improved if this spectral space were used, as Fukinuki has proposed, to increase the horizontal luminance resolution rather than to maintain the theoretical vertical chrominance resolution. The latter is low in present-day cameras and, in any event, cannot be used effectively because of the very low horizontal chrominance resolution. Increased cross color on conventional receivers is to be expected with this technique, though Fukinuki reports that this is a small effect.

2.3 Separation of Camera and Display Parameters from those of the Channel

There are a number of important objectives that can be realized more readily, both in improving NTSC and in developing entirely new systems, by separating the scanning standards of the camera and display from those of the channel. [22] One is the elimination of the defects related to interlace. Another is the improvement of the spatiotemporal resolution of the entire system. Both require explicitly controlling the frequency response of the transducers rather than relying on parameters that are an accident of their natural characteristics and of the channel scanning standards. Such explicit control requires that this response be defined in terms of a group of coefficients - certainly not less than 2 or 3 along each axis. At a bare minimum, the sampling density of the transducers must therefore be at least double that of the channel, per axis, or eight times the overall data rate. It is not clear that even this is enough if any reasonable precision is to be obtained, but at least it is a start.

2.3.1 Kell Factor

In analog television systems, it has long been a problem to equate the effective vertical and horizontal resolution, one direction being sampled and the other being rendered in a continuous manner. In the horizontal direction, it has been customary to assume two resolvable elements per cycle of bandwidth, in accordance with the sampling theorem. Equivalent resolution in the vertical direction is found by multiplying the number of lines by a factor less than one, justified on a number of shaky theoretical and experimental grounds. A recent study has shown that the Kell factor depends on the pre- and postsampling filters and can be very close to unity. [23] One of the advantages obtainable with high-rate cameras and displays is the achievement of a high Kell factor.

2.3.2 Temporal Performance

The role of the camera in degrading images of moving objects was referred to above. With the freedom to implement a variety of presampling filters that is provided by a high-rate camera, the trade-off between blurring and temporal aliasing can be made on a rational basis, rather than as a happenstance of integration time. For example, in some cases it may be preferable to maintain the sharpness of moving objects even at the cost of introducing stroboscopic effects, whereas in other cases the reverse may be true. In general, temporal performance should be improved when the output of a high-rate camera is properly prefiltered before subsampling to obtain the signal to be transmitted.

2.3.3 Interlace

A high-rate display, for example 525 or 1050 lines progressively scanned, can positively eliminate interline flicker, even without a high-rate camera. However, if simple-minded linear interpolation is used, some loss of sharpness results. In addition, some artifacts at moving sharp edges may develop. Rather good systems have been developed in which motion- *adaptive* interpolation is used, temporal in the stationary portions of the image, and spatial (vertical) in the moving areas. [24] Ideally, the interpolation should be motion- *compensated*, i.e., along the optical flow lines, since in this way, moving objects could be kept from blurring. [25] It is also possible that fully optimized fixed interpolation, as has been developed for graphic arts applications, would serve well here. [26]

With respect to vertical motion artifacts, the disappearance of half the scan lines at certain speeds is also eliminated simply by display up-conversion, as the line interpolation process is often called. Vertical-temporal aliasing, as discussed above, is primarily a function of the camera, and can be eliminated by appropriate prefiltering. Therefore, its amelioration probably requires a high-rate camera.

It should be noted that interlace causes few problems in the presence of low vertical camera resolution.[7] In that case, the use of up-conversion to eliminate a nonexistent problem may, in fact, reduce rather than improve the image quality, especially if the up-conversion itself is less than perfect. The villain, in this case, is not the up-conversion, but the low resolution, which wastes channel capacity.

3. New Systems

The advantages and disadvantages of a revolutionary approach mirror those of the approach based on improvements in NTSC. From the long-term point of view, there is a significant economic advantage in going to an entirely new system, in addition to the fact that it would be possible to have arbitrarily high quality. Surely, if it became necessary to replace every piece of TV equipment in the world, both consumer and professional, we would all end up richer, not poorer. That route, after all, has been the history in many other fields. The creation of new industries, such as the automobile industry, is the path to wealth, not poverty. The problem is avoiding the dislocations that inevitably accompany the phasing out of products, companies, and industries.

From the technical standpoint, almost any new system can be made backward compatible by using a two-channel scheme, in which one channel is NTSC.[8] This, however, limits the efficiency of new systems. In particular, it prevents the development of any system that achieves theater quality in a 6-MHz analog channel, a possibility that we believe to be very real.

If the political obstacles can be overcome, the main advantage of a noncompatible system, in our opinion, is that its long-term economic future is much sounder than that of any compatible solution. There is no evidence that a large number of viewers will buy expensive receivers when they can see the same programs, albeit at lower quality, on their existing equipment.[9] Therefore, any compatible system runs the real risk of never developing the audience required to provide the economic justification for the production of high-definition programs on a large scale. A noncompatible system, on the other hand, requires that new receivers be used to view the new and presumably more desirable programs, especially if NTSC broadcasting is definitely to be stopped on some particular date, announced several years in advance. Since the new receivers could be made multistandard at moderate additional cost, there would also be an incentive for their purchase during the period when most of the programming were still NTSC.

From the standpoint of economic justice, this approach does not put an unreasonable burden on anyone. After all, no guarantee of permanent life is ever provided with any product, no matter how expensive or how durable. A TV receiver that cost a few hundred dollars and has lasted five years has been a much better investment than many other household appliances. We regularly trade in automobiles with a lot of life left, at a large loss, so that we can have new ones that provide only marginally better transportation. Perhaps we can learn to do the same with TV, especially as this change would be once in a lifetime, and the new sets could also receive the old signals.

3.1 How Good Should a New System Be?

At the very least, a new system should have a technical quality sufficiently superior to that of the current systems, as perceived by the mass of viewers in their homes, so that it would make sense to change. Some hold the view that this means at least twice the vertical luminance resolution[10] and a higher aspect ratio. Much higher chrominance resolution is mandatory, especially in the face of higher luminance resolution. Greatly reduced spatial aliasing without attendant blurring would be highly desirable. It goes without saying that digital stereo sound is an absolute requirement of any new TV system. Although there is little discussion of temporal issues, this author would like to see

---

[7]If the screen is viewed closely enough so that the scan lines are clearly resolved, interline flicker will be seen even with a blank field, a problem that requires display up-conversion for correction.

[8]A somewhat tongue-in-cheek proposal is to have one NTSC channel and one PAL channel, the combination becoming HDTV! With advanced signal processing and some tolerance of reduced performance on existing receivers, this is not entirely out of the question.

[9]There is, on the other hand, copious evidence that viewers are willing to pay substantial amounts of money for programs that they want to see, even when many other programs are available free.

[10]This specification is indefinite, since it is not apparent whether the comparison is to the vertical resolution achieved with NTSC on standard receivers or on properly up-converted displays, with signals originating in a high-line-rate camera. See the Fink articles mentioned in the bibliography.

smoother motion, higher spatial resolution of moving objects, and substantial absence of interline flicker. Of course, since any new system should produce the highest possible quality for the given channel capacity, it would naturally use many of the techniques discussed above for making NTSC more efficient.

It should be borne in mind that the system designer has little control over the viewing conditions. The impact of a display is greatly affected by the angle subtended at the viewer. The angle may well be different when the audience settles down to watch a spectacle on a large screen as compared to casually observing the news on a small set over breakfast. The large-angle display requires much more spatial resolution; for pictures viewed at 10 times image height, we already have enough. Thus there is some uncertainty in the following discussion.

3.2 Aspect Ratio

The motion picture industry's great weapon of the 50's against the inroads of television was the wide-screen film. [27] Most of the most successful films of those years were made in Cinemascope (*The Robe, Bridge on the River Kwai*) at 2.35:1, a format never intended to be jammed into the 4:3 "keyhole" of the small screen. Other wide-screen formats such as Cinerama (*This is Cinerama*), Vistavision (*The Ten Commandments*), and Todd-AO (*Around the World in Eighty Days*), had more fleeting popularity, but were of undoubted visual impact, as is IMAX today. [28] Two popular current formats are Panavision's anamorphic 2.35:1 (2:1 squeeze in the camera) and "flat" 1.85:1. In the face of this history, no observer tests are needed to determine the preferred aspect ratio - for entertainment productions such as drama or sports, the wider the better. For productions such as news and interview programs, as well as game and educational shows that are specifically designed for the small screen, 4:3 is just about right.

One thing that should be perfectly obvious is that thoughtful directors will turn out programs that "work" best in the intended medium. If these productions are then tested in a variety of presentations, they will be preferred in the format for which they were made.[11] This presents a serious problem when material is prepared to be used alternatively in the theater or for television, as is often the case under present-day economic conditions. If a production is to look right on TV, it cannot possibly take full advantage of the wide screen of the theater unless the director takes very special pains, probably shooting certain scenes two ways. It is therefore quixotic to advocate, say, a 2.35:1 aspect ratio, which would permit much greater visual impact. However, if we are ever to break away from the 4:3 strangle hold, we should pick as wide a screen as possible, while still getting some kind of consensus to use it. In view of its popularity in the film industry, 1.85:1 is certainly a good candidate, and the conversion to 4:3, while difficult, is not impossible.

3.3 Transcodability

Since NTSC, PAL, SECAM, and 24-fps film will be with us for a very long time, one of the most important issues in the design of any new system is the ability to be interchanged with all of these with high quality and acceptable cost. Historically, the film transcoder has proved very difficult, at least for 60-Hz systems. Since film provides much of the programming material for TV broadcasting, and the making of films by means of HDTV has been stated to be one of its important applications, this may be the most important problem.

If a TV system were to be designed only for making films, it would be perverse to use anything but 24 fps, progressively scanned. This would produce films much like those shot in a film camera, but with the greater speed and convenience of electronic production. Up-conversion to 48 or 72 fps for viewing would be simple.[12] It is also clear that if the audience were to be satisfied with today's motion rendition when film is shown on television, these same 24-fps progressive productions could easily be converted to 60-Hz interlace (or 50-Hz interlace by 4% speedup) for broadcast in NTSC or PAL. Line-rate conversion is rather straightforward for any ratio.

A more difficult problem is presented if one wants to transcode with better motion rendition than normally seen from 24-fps originals, either in the motion picture theater or on television. Here

---

[11]In our Audience Research Facility, we carried out one test of preferences for aspect ratio in monochrome photographs. Not surprisingly, portraits were preferred in "portrait" mode and landscapes in "landscape" mode. [29] I have no doubt that this is true for TV as well.

[12]Editing would probably be even easier with a frame synchronizer, fed directly by the 24-fps signal, but supplying 525-line, 60-Hz interlaced signals to the monitor.

the problem is not only transcoding but also capturing the motion adequately in the first place, a job not done very well at present by any motion picture or TV system.

3.4 Motion Rendition[13]

Although they do not use precise words to describe what they see, nonprofessional viewers of images do notice defects in SNR, sharpness, tone reproduction, and color rendition. [30] However, they are remarkably oblivious to motion defects, even of the most egregious sort. There has been very little serious study of motion perception; perhaps when there is, this situation may be clarified. It is also true that observers can be trained to notice motion problems, and when they are sufficiently sensitized to the phenomenon (as is this author), they may regard it as an important part of overall image quality. Just as interline flicker is not apparent to lay observers until they see an interlaced display alongside one with progressive scan, it may well be that, when good motion rendition is directly compared with poor, the difference will be obvious.

In the author's opinion, good motion portrayal is a vital part of making the TV viewing experience realistic. Unfortunately, experimental work so far indicates that there is no practical frame rate that gives flawless results. [31] However, we can do much better than we do now, especially when 24-fps film is used for 60-Hz TV.

3.4.1 Motion in Television

The main problem involves eye tracking. If camera and eye are stationary, stationary objects are correctly rendered and integration time blurs moving objects for both. If the camera is tracking a moving object at which the eye is staring, again the results are fairly good, since similar blurring due to integration occurs in the eye and the camera for the portion of the scene not being tracked. In both cases, the portion of the image that is stationary on both retina and camera focal plane is rendered well. The portion in relative motion may show some stroboscopic effects, depending on the camera exposure time, but these often will not be prominent because that area is not the primary object of attention. Of course, high-rate cameras and displays, combined with the appropriate filters, help a good deal, but the rendition is fairly good even without them.

If eye and camera are not tracking in the same manner, then excessive blurring and/or stroboscopic motion are likely to take place. For example, when the eye is tracking an object moving within the TV image, it sees the blurring resulting from camera integration time. If the latter is shortened by means of a mechanical shutter or the equivalent, this blurring is reduced but the stroboscopic effect in the background is increased. Likewise, when the eye is fixed and some important object starts to move, during the period when the eye is beginning to track, very strong stroboscopic effects may occur, in the form of multiple images. These effects are not completely removed even at 120 fps. [32] There appears to be no simple way to reduce them in new systems, except to operate at as high a frame rate as possible, and to perform appropriate pre- and postfiltering in connection with high-rate cameras and displays.

3.4.2 Motion in Film

Typically, 24-fps film is exposed with a 50% duty cycle,[14] about half the duration used in a standard TV camera. Therefore, images of moving objects will be less blurred, which is good if the object in question is being tracked by the observer within the scene, but bad if the object is also moving relative to the retina. The problems are exacerbated by the display, in which, to raise the large-area flicker rate, each frame is projected twice, or, occasionally, three times. Now, in the tracking case, since the eye moves at the average speed of the image, double (or triple) images are clearly seen. This is more apparent with sharp-edged objects such as moving titles, but it is present nearly all the time in ordinary scenes and very obvious once pointed out. When the movie camera is tracking objects, a similar effect takes place in the background, and, depending on the subject matter, gives rise to spectacular stroboscopic effects often completely ignored by viewers.

When 24-fps film is used on 60-Hz TV,[15] the difference in frame rates is smoothed over either

---

[13]This section is based on discussions with Andrew Lippman and Stephen Hsu.

[14]This is also referred to as a 180-degree shutter.

[15]In 50-Hz countries, the film is either shot at 25 fps rather than 24 fps, or is sped up by 4%. Motion rendition is similar to that seen in the theater.

by camera integration, which produces multiple images if these would be seen in the movie theater, or by the 3/2 pulldown method. In the latter method, successive film frames are used for 2 and 3 fields, respectively, producing a 12-Hz temporal component, and extremely jerky motion rendition.[16] As we shall see below, it is possible to do much better.

3.5 Spatiotemporal Resolution

If the conventional wisdom is followed in the choice of resolution and aspect ratio, we need about 1000 lines/frame, 1700 samples/line, and 30 fps, for a total luminance data rate somewhat more than 50 Megasamples/sec. This leads to bandwidth consumption on the order of that required by the NHK HDTV system. On the other hand, it is now recognized that spatial and temporal bandwidth may be traded off, reducing response for components simultaneously high frequency in all directions. In MUSE, for example, only stationary objects have full spatial resolution; moving objects have only 1/4 the area resolution. [33] In Glenn's system, fine spatial detail has only 1/4 the temporal resolution. [34] While both of these methods can be thought of as compression systems, it is more meaningful to describe them as systems that attempt to use the available channel capacity in the psychophysically most advantageous manner.

One method of trading off resolution along the various axes is to use an appropriate spatiotemporal sampling pattern. [35] Each such distinct pattern leads to a distinct shape for the alias-free Nyquist baseband, which defines the resolution.[17] The offset pattern, shown in Fig. 6, makes what appears to be an appropriate compromise. Offset patterns have at least some response at frequencies 41.4% higher than the Cartesian pattern. For example, 750 lines/frame would give a peak vertical response of 1061 lines, which, with proper filtering and up-conversion, would extend the vertical MTF to more than 500 cph, or 25% higher that normally observed on the NHK 1125-line system. Likewise, the temporal response would extend to 15 Hz with only 21 fps.

3.6 Composite or Component?

As discussed above, we now know how to operate a composite system with essentially no cross effects. It is therefore not necessary to go to a component system, such as one of the various MAC schemes, to solve this problem. [36] In fact, the composite system has an unintended advantage, in that the portion of the 2- or 3-d spectrum that must be sacrificed to make it work does not contribute very much to luminance quality. Undeniably, however, the component schemes are simpler, which is counterbalanced, in most of those so far proposed, by the fact that they are less efficient, i.e., they take more channel capacity for the same resolution.

In view of the fact that the desired shape of the Nyquist baseband can be obtained by using the appropriate offset sampling pattern together with the corresponding filters, it would appear that the better choice for new systems would be of the component type,[18] with filtering and sampling to obtain a baseband in which the higher frequencies along any one axis have a lower response along the other axes. This implies that high-rate cameras and displays are used, with 3-d filtering. This, in turn, implies the presence of frame and line stores at both transmitter and receiver, so that the channel frame rate need not be high enough to avoid either large-area or interline flicker.

3.7 Some Example Systems

We now briefly discuss a few sample systems. None of these is seriously proposed as a standard of any kind; each is mentioned only for the purpose of illustrating specific points. All new systems are to use frame stores and high-rate cameras and displays. The question of aspect ratio is finessed by speaking only of the total number of samples per frame; these may be rearranged as desired.

3.7.1 A System for the 6-MHz Analog Channel

---

[16]This author finds it striking that this is tolerated by American TV professionals. In contrast, the insistence of European TV interests on near-perfect rendition in the NHK HDTV-to-PAL transcoder seems to imply a quite different view of TV quality.

[17]These basebands are not unique. However, there is a "natural" baseband, of maximum symmetry, for each sampling pattern. Except for the Cartesian pattern, they all have a temporal frequency response that varies continuously with spatial frequency, and vice versa.

[18]It follows that the composite technique should find its principal application in improving existing systems, and not in the design of new ones.

Assuming that it would be harder to change channel allocations than TV system standards, it is of interest to see just how good an image can be transmitted in the existing terrestrial channels. We note that only a 4.2-MHz signal can at present be used in a 6-MHz channel, and that the horizontal definition is reduced by more than 15% by retrace time. The first step, therefore, is to eliminate vestigial-sideband transmission and the separate sound carrier, and either to eliminate the retrace time or use it for something important. A single carrier, in the middle of the band, could carry two independent 3-MHz channels by double-sideband quadrature modulation. To minimize adjacent-channel interference, the outer extremities of the band should carry only detail information (rather than any low-frequency data), perhaps by having each of the two signals carry information for half the lines.

Assuming that some advantage in motion portrayal will ensue from use of proper filtering, we should be able to use no more than 18 fps. If the channel, now a full 6-MHz, is good for 12 Megasamples/sec, we can have 667,000 samples/frame, or at least three times that used at present. We still must provide for color and audio. Ten percent of the capacity should be sufficient for the former and five percent for the latter, still leaving a picture very much better than seen at present.

3.7.2 A System for 216 Megabits/sec.

CCIR 601, the standard for digital transmission of current TV signals, provides an image substantially superior to that of the best analog transmission, even without using any special advanced techniques. Since transmission facilities and, perhaps more importantly, digital VTR's are being provided for this standard, it is interesting to see what can be done to raise image quality using some of the methods mentioned here.

Digital transmission permits the use of two additional techniques - spatially offset sampling as advocated by Wendland, [37] and relatively coarse quantization of samples representing the spatial high frequencies. [38] In the latter scheme, each frame is divided by 2-d spatial filters into highs and lows.[19] The three-color lows are coarsely sampled and finely quantized, and the monochrome highs are finely sampled and adaptively coarsely quantized. Using 24 fps and allowing a full 2.5 Mb/s for digital audio gives nearly 9 Mb/frame. If the lows have 1/3 resolution in each direction, each block of 9 samples requires 27 bits for the color lows using 9 bits/sample, giving a SNR in the blank areas of 59 db. The highs take 4 bits/sample plus 4 bits/block of adaptation information for a total of 7.44 bits/sample, allowing nearly 1.2 Megasamples/frame. When these are spatially offset in successive fields as shown in Fig. 6, and then interpolated up to double resolution for display at 72 fps, progressively scanned, we would expect to have images superior to those of the NHK HDTV system with respect to flicker, motion smoothness, sharpness of moving objects, spatial resolution, and SNR.

3.7.3 A Very High Quality TV System (VHQTV)

Hitachi has made an experimental 500 Mb/s DVTR and Sony is said to have made a similar machine of about 1 Gb/s capacity. If such were really available, with the cameras and displays to match, what kind of a system would be most appropriate? It is hard to say. Surely, the 216 Mb/s system described in the preceding section would satisfy the most demanding consumer application. However, there may be a place for even higher performance for certain kinds of public displays. Two such systems have been shown in the theater. One is IMAX, 24 fps with a very large film area (48x69 mm), and the other is Showscan, 60 fps with a more modest film area, but still much larger than standard 35 mm. Without extensive study, it would not be possible to determine the best way to use the additional capacity. To hedge the bet, one could increase the frame rate modestly, to 36 fps. What to do about the spatial rendition depends largely on whether the additional 50% of spatial data is to be used to increase the sharpness or the size of the final image. If the sharpness is to be increased, then adding a third channel, at perhaps 2 bits/sample, would double the area resolution. If the image size is to be increased, then it would go up only in proportion to the data rate - 50% on an area basis.

4. Motion Compensation

It has been known for a long time that successive frames must be very much alike to achieve an illusion of continuous motion. This has been the basis of a number of coding and noise-reduction proposals. [39] A more advanced concept is to take advantage of knowledge of the actual motion, or

---

[19]There is a 3-d extension of this system, which is somewhat more efficient, but its quality has not yet been completely proved out.

displacement, of corresponding points in successive frames. Such knowledge can be used for noise reduction without blurring, predictive coding, and adaptive interpolation. Like other adaptive techniques, motion-compensated processing is always better than nonadaptive processing, which is, after all, just a special case, *except when the motion estimation is wrong.*[20] Present-day estimation methods are good enough for important practical applications. The major obstacle is not the accuracy, but the speed and cost of the associated calculations.

Evidence that motion compensation is useful is given by three recent demonstrations. NHK uses motion-compensated interpolation for MUSE [40] and for the HDTV-to-PAL transcoder. [41] The BBC Research Laboratories has shown a 4:1 compression system, called digitally assisted TV (DATV), using subsampling and motion-compensated interpolation. [42] The first two systems, which are quite successful, use a rather crude form of this processing. DATV works well where the motion estimates are accurate.

4.1 Motion Estimation at Receiver or Transmitter?

An important issue is whether the motion information is to be calculated at the transmitter and sent in a side channel or whether it is to be calculated at the receiver. Using a side channel for auxiliary data effectively broadens the signal description to include velocity information. There is no principle of signal processing that shows that the optimum signal description consists only of a sequence of sample values. In describing the focal plane light intensity function (what we prefer to call the *video function,* to distinguish it from the *video signal),* it is quite possible that a more efficient (compact) description would include some motion information, which corresponds, more or less, to derivative information. In addition, if a high-rate camera is used, there is more data available at the transmitter to calculate the motion accurately. Finally, since there is one transmitter and many receivers, good system design would put the major burden of this calculation at the sending end. It is possible that a simple calculation could be done at the receiver, with the side channel being used to transmit supplementary motion information.

4.2 Motion-Compensated Temporal Interpolation

An important application of this technique is to get a more advantageous trade-off between spatial and temporal resolution. For example, in the 24-fps systems described previously, comparable performance could very likely be obtained at 12 fps. For the same data rate, this would permit doubling the area spatial resolution, for a notable improvement in overall image quality.

Another application is the conversion of 24-fps film to 60-Hz TV. We have done some experiments on this subject in our laboratory with promising results. Especially with computer-generated images having the full spatial resolution permitted by the standards, the elimination of jerky motion is quite spectacular. With natural (camera) images, the improvement, although noticeable, is much less, a fact we attribute mainly to typically poor camera performance.

The performance at 24 fps was so good, we have tried one experiment at 12 fps, using only motion information derived from the final 12-fps signal. (In a practical case, we would have the original camera sequence available and could do a better job of motion estimation.) The results were, at the very least, promising. There is a strong possibility, in our opinion, that 12 fps will be shown to be a completely adequate frame rate when motion compensation is used for final viewing.

4.3 Motion-Compensated Noise Reduction

Noise reduction is of commercial importance. Although it is true that modern TV cameras, with adequate illumination, produce signals that need no noise reduction, multigeneration analog recording, ENG equipment, low-light-level picture taking, and satellite transmission all produce signals that have greater-than-optimum noise levels. The present generation of frame-recursive motion-adaptive noise reducers, all descendants of ideas proposed in the early seventies, [43] trade off noise reduction against blurring of moving objects. The effective time constant of a temporal low-pass filter, using feedback around a single-frame memory, is adjusted in accordance with the amount of frame-to-frame motion, imputed from the measured difference between the same points in successive frames. A great deal of smoothing is used in stationary areas and less and less in the presence of more and more motion. In some cases, a "slow switch" in used, in which the time constant is based on an area

---

[20]Noise is the enemy of accurate motion estimation. Recursive estimation, in which the noise is reduced, permitting more accurate estimation as the estimate converges, shows some promise.

measurement rather than on a point-by-point measurement.

These noise reducers tend to blur out low-contrast moving detail and they fail to reduce noise in moving objects. When the latter are tracked by the viewer, the noise visibility is hardly reduced. Both of these defects can be cured by integrating in x,y,t space, not parallel to the time axis, but along the motion vector. If smooth motion persists long enough to be tracked (more than about 200 msec.), it would appear that the noise could be reduced at least 6 db, and in most cases, 12 db or more, with negligible reduction in sharpness. [44]

5. Conclusion and Acknowledgements

After a discussion of the limits to received image quality in NTSC and a review of various proposals for improvement, it is concluded that the current system is capable of significant increase in spatial and temporal resolution, and that most of these improvements can be made in a compatible manner. Newly designed systems, for the sake of maximum utilization of channel capacity, should use many of the techniques proposed for improving NTSC, such as high-rate cameras and displays, but should use the component, rather than composite, technique for color multiplexing. A preference is expressed for noncompatible new systems, both for increased design flexibility and on the basis of likely consumer behavior. Some sample systems are described that achieve very high quality in the present channels, full "HDTV" at the CCIR rate of 216 Mb/s, or "better-than-35-mm" at about 500 Mb/s. Possibilities for even higher efficiency using motion compensation are described.

The Advanced Television Research Program at MIT has been supported since June 1983 by the members of the Center for Advanced Television Studies, a US television industry group. At various times, ABC, CBS, NBC, PBS, Time Inc., Ampex, Harris, Kodak, Tektronix, RCA, 3M, and Zenith have been members of the group. Their sponsorship is gratefully acknowledged.

Many of the ideas discussed in this paper are the outgrowth of discussions with sponsors, colleagues, and students, whose contributions have been essential to the work done on this program. The opinions expressed herein, however, are those of the author alone, and not of MIT or the sponsoring companies. The work has been carried on in the Research Laboratory of Electronics and the Media Laboratory.

Appendix. The Spectrum of the Composite Signal

We call the light intensity in the focal plane of a camera the *video function* and note that it is a function of x, y, and z, z standing for time. Within the borders of the image and within an arbitrary time interval, the function can be represented by a triple Fourier series. The latter, being periodic in the width A, height B, and time interval C, thus replicates a triple infinity of identical functions outside the A, B, C box.

A typical term of the Fourier series is

$$\exp[2\pi j\,(mx/A + ny/B + pz/C)\,]$$

where m, n, and p are the number of cycles per image width, height, and duration, respectively. We call these terms spatiotemporal harmonics. To find the signal that results from scanning, we note that repetitive scanning downward within the box is equivalent to unidirectional scanning in $x, y, z,$ space according to the expressions

$$x = Af_h t;\quad y = -Bf_v t;\quad z = Cf_t t.$$

Thus, the video signal is also represented by a Fourier series with a typical term

$$\exp[2\pi j\,(nf_h - mf_v + pf_t)\,t],$$

where each spatiotemporal harmonic of the video function corresponds to one temporal harmonic of the video signal.

In the NTSC system, the color subcarrier frequency is related to the horizontal and vertical scanning frequencies by the following expressions:

$$f_{sc} = (455/2)f_h = (455/2)(525/2)f_v.$$

The positive frequency color subcarrier can thus be thought of as resulting from the scanning of a spatiotemporal harmonic in which

$$n = 227, \ m = -131, \ f_t = 15 \ (Hz), \ and \ n = 228, \ m = 131, \ f_t = -15,$$

while the negative subcarrier frequency results from

$$n = -227, \ m = 131, \ f_t = -15, \ and \ \ n = -228, \ m = -131, \ f_t = 15.$$

Note that for a harmonic to be real, there must be a pair of exponential (complex) harmonics in diametrically opposite octants.

The subcarrier harmonics are located in four of the 8 octants in 3-d frequency space, as shown in Fig. 1. Each such harmonic is surrounded by sidebands representing the chrominance signals. Since the scanning patterns for chrominance and luminance are the same, the "natural" bandwidths in all three directions are also the same. The horizontal chrominance bandwidth is restricted by the temporal filters used before and after combining the components to form the composite signal, but there are normally no provisions for limiting the vertical and temporal bandwidths. As shown in Figs. 2 and 3, if cross effects are to be controlled, it is required to shape both the luminance and chrominance spectra in all three dimensions. Most proposed techniques result in substantially less bandwidth for chrominance than for luminance.

References

1. W.F.Schreiber, "Psychophysics and the Improvement of Television Quality," SMPTE J., 93, 8, Aug. 1984, pp. 717-725.
2. D.Lemay, "Interpolation Temporale avec Compensation du Mouvement," MSc Thesis, INRS-Telecommunications, 5/85.
3. A.Roberts, "The Improved Display of 625-Line Television Pictures: Adaptive Interpolation,: BBC Res. Rept., May 1985.
J.L.E.Baldwin, "Enhanced TV - A Progressive Experience," SMPTE J. 94, 9, Sept. 85, pp. 904-913.
K.H.Powers, "Techniques for Increasing the Picture Quality of NTSC Transmission in Direct Satellite Broadcasting," IEEE J. Sel. Areas Comm. 3, 1, 1/85, pp 57-64.
4. R.Turner, "Some Thoughts on Using Comb Filters," IEEE Trans. Consumer Electr., CE-23, 3, 8/77, pp. 248-256.
S.J.Auty et al., "PAL Color Picture Improvement of Television Quality," SMPTE J., 87, 10, 10/78, pp 677-681.
5. T.Fujio, et al., "HDTV," NHK Tech. Monograph, 32, June 1982.
6. J.L.E.Baldwin, "Analog Components, Multiplexed Components, and Digital Components -- Friends or Foes?," SMPTE J. 92, 12, Dec 1983, pp. 1280-1286.
7. E.F.Brown, "Low-Resolution TV: Subjective Comparison of Interlaced and Non-Interlaced Pictures," Bell Sys. Tech. J., 46, Jan. 1967, pp. 199-232.
8. T.G.Schut, "Resolution Measurements in Camera Tubes," SMPTE J., Dec. 1983, pp. 1270-1293.
9. I.C.Abrahams, "The Frequency-Interleaving Principle in the NTSC Standards," Proc. IRE, 42, 1, 1/54, pp. 81-83.
10. G.T.Flesher, et al., "The General Theory of Comb Filters," Proc. NEC, 17, 1961, pp. 81-94.
11. F.Gray, "Electro-optical Transmission System," USP 1,769,920, 4/30/29.
12. I.C.Abrahams, "Choice of Chrominance Subcarrier Frequency in NTSC," Proc. IRE, 41, 1, Jan. 1954, pp. 79-81.
13. M.W.Baldwin, "The Subjective Sharpness of Simulated TV Pictures," Proc. IRE, October 1940, pp. 458-468.
14. J.Fourier, "Theorie Analytique de la Chaleur, "Didot, Paris, 1822.
15. P.Mertz and F.Gray, "A Theory of Scanning and its Relation to the Characteristics of the Transmitted Signal in Tele- photography and Television," Bell Syst. Tech. J., 13 July 1934, pp. 464-515.
16. N.W.Parker, "Color TV Signal Separation System," USP 3,542,945, 9/11/67.
17. J.O.Drewery, "The Filtering of Luminance and Chrominance to Avoid Cross-Colour in a PAL Colour System," BBC Engineering, 8, 39, 9/76, pp. 8-39.
E.Dubois, et al., "The Three-Dimensional Spectrum and Processing of Digital NTSC Color Signals," SMPTE J., 91, 4, Apr. 82, pp. 372-378.
18. E.Dubois and P.Faubert, "Two-Dimensional Filters for NTSC Color Encoding and Decoding," IBC 86.
19. Y.Faroudja, "Adaptive Comb Filtering," USP 4,179,705, 3/13/78.
J.Rossi, "Comb Filter for TV Signals," USP 4,050,084, 7/14/76.

A.Kaiser, "Comb Filter Improvement with Spurious Chroma Deletion," SMPTE J., 86, 1, Jan. 1977, pp. 1-5.

20. T.Fukinuki and Y.Hirano, "Extended Definition TV Fully Compatible with Existing Standards," IEEE Trans. Commun., 32, 8, Aug. 1984, pp. 948-953.

21. T.Fukinuki, Y. Hirano and H. Yoshigi, "Experiments on Proposed Extended-Definition TV with Full NTSC Compatability, SMPTE J. 93, 10, Oct. 84, pp. 923-929.

22. B.Wendland and H.Schroder, "Signal Processing for New HQTV Systems," SMPTE J., 94, 2, Feb. 1985, pp. 182-189.

23. S.Hsu, "The Kell Factor: Past and Present," SMPTE J. 95, 2, Feb. 86, pp. 206-214.

24. R.N.Jackson and M.J.J.C.Annegarn, "Compatible Systems for High-Quality Television," SMPTE J. 92, 7, July 83, pp. 719-723.

25. D.M.Martinez, "Model-Based Motion Estimation and its Application to Restoration and Interpolation of Motion Pictures," Ph.D. Thesis, M.I.T. EECS Dept., August, 1986.

26. W.F.Schreiber and D.E.Troxel, "Transformation Between Continuous and Discrete Representations of Images: A Perceptual Approach," IEEE Trans. Pat. Anal. & Mach. Intel., 7, 2, March 1985.

27. "Widescreen," Special issue of *The Velvet Light Trap*, 21, Summer 1985, Madison, Wisconsin.

28. W.C.Shaw, et al., "IMAX and OMNIMAX·Theatre Design," SMPTE J., 92, 3, March 1983, pp. 284-290.

29. H.D. MacDonald, "The Relative Quality of Television and Motion Pictures, MS Thesis, M.I.T. EECS Dept., Jan. 1985.

30. R.W.Neuman et al., "Report of Preliminary Focus Group Interviews," MIT/ATRP report, 1984, unpublished.

31. G.Tonge, Meeting Report, "Les Assises des Jeunes Chercheurs," Rennes, France, 9/9/85.

32. S.C.Hsu, "Motion-Induced Degradations of Temporally Sampled Images," SM Thesis, M.I.T. EECS Dept., June 1985.

33. Y.Ninomiya, Y.Ohtsuka, and Y.Izumi, "A Single Channel HDTV Broadcast System - The Muse," NHK Lab. Note 304, Sept. 1984, pp. 1-12.

34. W.E. Glenn, "Reduced Bandwidth Requirements for Compatible HDTV Transmission," Presented at IGC HDTV Conf., New Orleans, Apr. 84.

35. Op cit ref 1.

36. D. Lowry, "B-MAC, An Optimum Format for Satellite TV Transmissions, "SMPTE J., 93, 11, 11/84, pp. 1034-1043.

H.Mertens and D.Wood, "The C-MAC packet System for Direct Satellite TV," EBU Tech. Rev., 200, 8/83.

37. Op cit ref. 22.

38. W.F.Schreiber and R.R.Buckley, "A Two-Channel Picture Coding System: II-Adaptive Companding and Color Coding," IEEE Trans. on Comm., Com-29, 12, Dec. 1981., pp 1849-1858.

39. J.E.Cunningham, "Temporal Filtering of Motion Pictures," Ph.D. Thesis, M.I.T. EE Dept., June 1964.

F.W.Mounts, "A Video Coding system with Conditional Replenishment," Bell Sys. Tech. J., 48, 7, Sept. 1969.

40. Op cit ref 33.

41. M.Sugimoto et al., "Second Generation HDTV Standards Convertor," 14th Int'l TV Symposium, Montreux.

42. BBC Research Laboratory, private communication.

43. D.Connor, Apparatus for Improving Video SNR," USP 4,040,221, 3/13/72.

J.Drewery and M.Weston, "Noise Reduction in TV Signals,: USP 4,058,836, 4/26/76.

A.Kaiser, "Noise Reduction System for Color TV," USP 4,064,530, 10/10/76.

44. E.Dubois and S.Sabri, "Noise Reduction in Image Sequences using Motion-Compensated Temporal Filtering," IEEE Trans. Comm.,73,4,7/84, pp 502-522.

Bibliography

D.G.Fink, "The Future of High-Definition Television: First Portion of a Report of the SMPTE Study Group on High-Definition Television, SMPTE J., 89, 2, Feb. 1980, pp. 89-94.

D.G.Fink, "The Future of High-Definiiton Relevision: Conclusion of a Report of the SMPTE Study Group on High-Definition Television," SMPTE J., 89, 3, March 1980, pp. 153-161.

G.J.Tonge, "The Sampling of Television Images," IBA, Experimental and Development Report 112/81.

G.J.Tonge, "Signal Processing for HDTV," IBA Report E8D, July 1983, USA.

D.E.Pearson, Transmission and Display of Visual Information, Pentech Press, 1975.

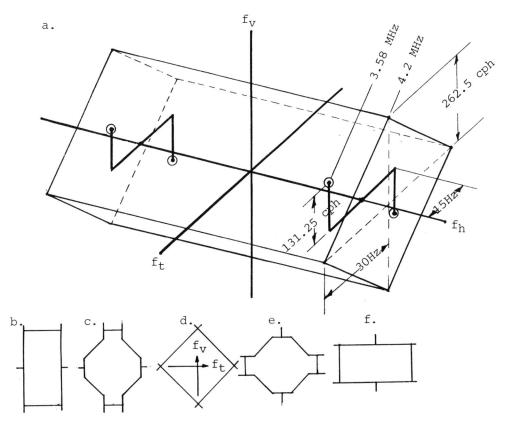

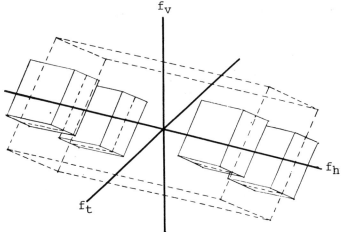

Fig. 1. *Location of the Subcarrier in the 3-D Frequency Spectrum.* The equivalent spatiotemporal subcarrier is located in four of the eight octants. The cylinder with diamond-shaped cross section is the largest symmetrical alias-free baseband possible for luminance with interlaced scanning. Other shapes of the cross section are possible as shown in Figs. 1b-f. The shape in Fig. 1b corresponds roughly to full-frame integration, while 1f corresponds roughly to single-field integration, as is more common in current TV cameras. The shape of Fig. 1a is believed to represent a better trade-off of vertical and temporal resolution.

Fig. 2. *Overlap of Chrominance and Luminance.* The chrominance spectrum is centered around the subcarrier in the same way the luminance spectrum is centered at the origin. The dotted lines show the largest possible alias-free luminance baseband while the four smaller diamond-cross-section cylinders show the chrominance with one half the vertical-temporal bandwidth. Obviously luminance must be made smaller to avoid overlap. Note that the horizontal extent of each component is limited by temporal filtering before and after combining them to form the composite signal. Conventionally, no specific vertical-temporal filtering is done at all.

171

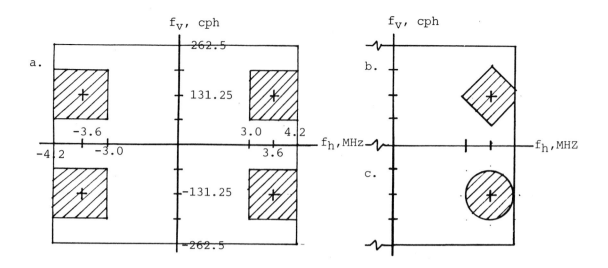

Fig. 3. *Separation with 2-D Filtering.* Ignoring temporal effects, the spectrum can be divided in the vertical-horizontal frequency plane as shown, totally eliminating cross effects. Other shapes can be used for chrominance, as shown in Fig. 3b,c. For the sake of simplifying the filters, some of the potentially usable spectrum space can be discarded. In Fig. 3a, the vertical chrominance resolution is only 65 cycles/picture height, or 1/4 that of luminance at low horizontal frequency. The vertical luminance resolution at high horizontal frequencies is also quite low.

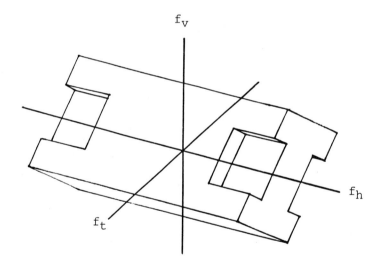

Fig. 4. *Separation with 3-D Filtering.* The luminance baseband is shown notched out to make room for chrominance. Many different allocations are possible. In this one, full spatial resolution is achieved for luminance and half spatial resolution for chrominance at zero temporal frequency, i.e., for stationary images, while both are reduced at higher temporal frequencies.

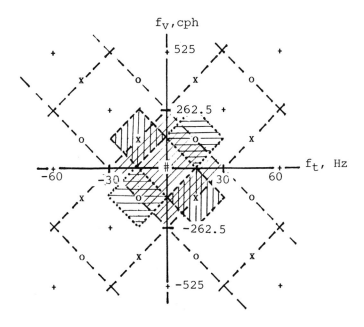

Fig. 5. *The Fukinuki Hole.* Here the spectra are shown in the vertical-temporal frequency plane. The baseband luminance, diagonally cross-hatched, is centered at the origin (#), with replicas centered at the harmonics of the line-scanning frequencies (+). The diamond cross section shown, which is believed to be optimum, provides maximum spatial resolution for stationary images. The color subcarrier and its harmonics are shown by the X's, vertically cross-hatched, with the chrominance spectrum extent chosen to be 1/2 of that of the luminance. With this choice, an equal space remains for the Fukinuki subcarrier at the O's, surrounded by the spectrum of an auxiliary signal, horizontally cross-hatched.

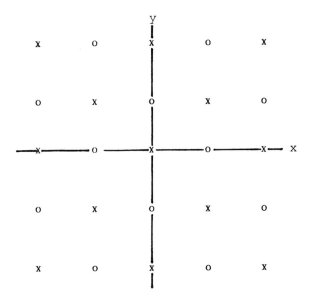

Fig. 6. *Spatiotemporally Offset Sampling.* The X's show the sampling grid in the even fields while the O's show the odd-field samples. With this pattern and the appropriate 3-d filters, the highest possible resolution is achieved along each axis in 3-d frequency space.

William F. Schreiber attended the New York City public schools and Columbia University, where he received the B.S. and M.S. in electrical engineering. In 1953, he received the Ph.D. in applied physics at Harvard University, where he was a Gordon McKay and Charles Coffin Fellow. Dr. Schreiber worked at Sylvania from 1947 and 1949 and at Technicolor Corp. in Hollywood, Calif. from 1953 to 1959. Since then he has been a faculty member at MIT, where he is now professor of electrical engineering and director of the Advanced Television Research Program. He was visiting professor of electrical engineering at The Indian Institute of Technology, Kanpur, India, in 1964-66, and at INRS-Telecommunications, Montreal, Quebec, 1981-82. Since 1948, Dr. Schreiber's major professional interest has been image processing. He has worked in graphic arts, including color processing and laser scanner design, in facsimile, and in television. This work has included theory and extensive practical applications. He is a Fellow of the IEEE, an Honors Award recipient of the Technical Association of the Graphics Arts, and has twice received the best paper award of the SMPTE *Journal.*

Evolution During Revolution

Guy Gougeon
CBC Engineering
Montreal, Canada

Introduction

The future of the TV broadcaster is today very unclear as the effects of new technologies such as the digitalization of current operations, the introduction of HDTV, and explosive growth in the number and kind of distribution channels are considered. The broadcaster is pulled forward inexorably by the seemingly insatiable requirements of viewers for more programs, better programs and higher technical quality, fuelled by new developments in consumer electronics. Innovation is needed to contain his production costs, to meet the viewers' demands for the quality and quantity and to retain a reasonable share of viewing time and of slots on the channel selector. The competition from other media such as cable, quasi-DBS, pre-recorded discs and tapes is strong. The broadcaster of today is faced with the dilemna of planning this future in the face of a plethora of competing but incomplete systems, many with confusing economic and political complications. At the same time, new uses for the large investment in plant and equipment owned by himself, his communications partners and his viewers must be evolved. There is no doubt that the need for innovation and major change is certain — the questions are concerned only with when and how. In this paper, the critical technical questions facing the broadcasting industry will be summarized and some possible future directions will be outlined.

Retrospective

A few years back, the broadcasters' task was relatively simple, being to produce, distribute and broadcast in the same standard with little competition. The principal problem was to build enough terrestrial transmitters to get the signal to the viewers. A little later, cable came along to assist in improving coverage and to hinder by adding more competing signals to the viewers channel selector. Then came satellites, first providing the "impossible" routes and later to revolutionize broadcast distribution patterns. Satellites next enabled the creation of "super-stations" and specialized services in conjunction with cable systems. Some viewers even installed their own receiving equipment to establish a de-facto C-Band DBS service — unplanned but quite durable. While the broadcaster now had much added competition and many viewers watched the signal only on cable, the broadcaster still had considerable technical, economic and political influence and could set the technical rules and programming norms to a large extent. The major technical decisions made were in regard to production equipment — broadcasting was comfortably composite everywhere outside his plant. The industry was not ready for next step.

The Components of the Revolution

In the space of a few years emerged the concept of Direct Broadcasting Satellites (DBS), followed closely by the demonstration of High Definition Television and strong efforts to introduce digital techniques into production and the viewer's receiver. In retrospect, the timing of these events could have been improved as DBS planning was based on composite coding and took little account of component transmission or of HDTV, while digital standards were developed just at the time when component operation was seen to be viable, but took no account of HDTV. Doubtless, some adjustments will have to be made. The broadcaster now faces challenges on several fronts: -

It is now clear that component-based production has definite advantages in quality, portability and cost but the areas to which it should be applied are less clear.

The introduction of all-digital facilities into the production system will occur over the next 3 years, beginning in post-production, where the benefits are potentially greatest. A concurrent rethinking of the production processes may be required.

The implementation of the digital transmission network proceeds rapidly terrestrially but this is not yet the case in the satellite network. The broadcasters network will then consist of a mixture of digital and analog sections with various interconnections. How can he plan a digital component network as a long-term goal, yet retain the necessary mixed facility composite network in the interim?

HDTV now seems economic for some production uses and as the technology matures, more sectors of production could be considered for its use. What is an appropriate introduction scenario and how does this relate to that for the introduction of digital television?

The appropriate technology and system for the delivery of HDTV or quasi-HDTV to the viewers' home is still a noisy debate and there is no agreement even of the definition of HDTV in this context. What new video services are desirable, given the current state of receiver development, economics and program resources?

The reality of TV program distribution is that the delivery to the viewer may be made by VHF or UHF terrestrial broadcast, SHF microwave, DBS (either de-facto or planned), cable, tape or disc. Each of these has a unique set of economic and technical advantages and disadvantages. Is it not time that the political and regulatory aspects be addressed so more optimal use of the currently competing resources can be achieved?

These questions are not independant and separable, as they focus on differing aspects of the production - distribution - delivery chain making a global solution difficult. As a first step it is useful to examine them individually and in the later panel discussion, the related effects can perhaps be explored to refine further the proposals for the moves from confortable composite to revolutionary components, in either low or high definition.

The Dilemma of Components

Where do components belong in the television plant? There is general agreement that the use of video components e.g. RGB or YUV can improve picture quality in production processing, recording and in applications where FM signal transmission is used, such as satellites and microwave links. Component operation also offers useful increases in color bandwidth for production processes, such as colour matte. It is also clear that composite NTSC is well suited for the limited bandwidth, AM-VSB terrestrial transmission system to the viewers, making very efficient use of this spectrum. New NTSC techniques, such as adaptive comb filters in decoders and complimentary pre/post filtering can eliminate many of the interface artifacts.

In the last few years, the newly introduced component systems have shown large improvements and are now adequate for production in the field and for many of the simpler activities in the studio. Improvements to NTSC will certainly keep it viable as a broadcast format for some time and we may even see the introduction of Dr. Fukinuki's proposal for super NTSC with luminance bandwidths of about 5.5 MHz, yelding very high quality pictures at home. Composite NTSC will remain, for some time in terrestrial broadcast, distribution and the areas related to them. Standards for component operation are now available allowing the development of production systems using this technology.

The dotted line between components and NTSC will move over a period of time, further and further out towards the receiver, but it will be many years before the current broadcast system disappears, as it is still capable of growth and because of the large investments in it by broadcasters and viewers alike.

CBC has already moved heavily into component equipment for news and information. This trend is expected to continue as component equipment improves and evolves allowing its use in entertainment and drama. We are also performing theoretical studies and computer simulations of the possible evolvement of the distribution network, considering both strictly compatible and transmission channel compatible improvements. Many studies and tests have also been performed on component-based distribution systems such as MAC, but there remains a number of question to be resolved in regard to its widescale use.

The Introduction of the All-Digital Studio

In the absence of digital delivery schemes, the all-digital studio becomes a production unit for a composite or MAC-based delivery system. Its principle advantages lying in its flexibility of processing and transparency of operation. A simple digitalisation of current production methods fails to capture these benefits effectively. What seems to be evolving is the "work-station" concept obtained by the linking of digital processing, graphics, digital disc and digital tape with good control software and artist interfaces. A new breed of artist must arise to use such equipment creatively. Are the program producers in broadcasting ready for this challenge that is now ariving on our doorstep and which will be fully usable in one or two years? It is our view that in 3 or 4 years beyond that, HDTV implementation of the same technology, in digital form, will be technically and economically viable also, complicating further the planning of new studios. CBC has taken an agressive position - building a development studio based solidly on the "digital-work-station" approach to gain the necessary technical and production experience in the shortest time and concurrently undertaking prototype productions in HDTV. Firm conclusions from this work will be available in 12 to 18 months to guide facility design. It is clear to us that a revolution in production is near at hand and will be pushed even further with the arrival of a studio camera based on CCD sensors and with the continued development of 3D graphics into real-time.

The Evolution of the Transmission Network

Current broadcasting networks are used both for transmission between studio centers and for distribution to viewers by terrestrial transmitters. They are complex networks made up of satellite, microwave and cable circuits, both temporary and permanent, with analog linkages. Frequently a number of videotape-based delay centers are included for time-zone correction. They have developed over a number of years and the configuration and use depend on the single-standard NTSC for all applications.

Over a period of years, this network must evolve to use digital links, many based on optical fibers, and will have to interface with the broadcaster in NTSC, components (both digital and analog) and HDTV (both in digital and analog forms). There may also be more than one level of service desirable. The linkages in the network must be at the digital level.

A number of principles can be established to guide the development of this new network.

- The broadcaster must define transmission building block modules of data that will work efficiently in the integrated services data network (ISDN), dedicated circuits, satellites and the current network. Transmissions can then be made by using one or more of these "modules" with suitable linking and coding to achieve the bandwidth and level of service required. For instance a single module could be used for ENG and two would be required to transmit NTSC.

- The network must be transparent eventually, without any transcoding by the carrier. In the short-term this may be difficult to achieve until digital facilities and switching become widespread.

- At switching points in the network, the associated "modules" must be kept together, though in the network their carriage on different circuit may have advantages in error management.

- We must agree on the sizes and formats of the modules so that switching becomes feasible, or at least predictable. It seems unlikely that a single format will be adequate but a closely-related family is certainly essential

- The need for customer-controlled switching, to improve network flexibility, has already been proven in the current network. It must be reviewed and expanded in the light of the ISDN and digital links that use several signal formats.

- The temptation must be avoided, to develop interim digital networks, based on single, inflexible schemes for the transmission of NTSC only. While this is certainly a short term priority and a need that will exist for some time, we must look forward to components and HDTV, and the evolution through a mixed network to a fully digital one.

Where does HDTV fit into the Plan?

There is some confusion, in regard to the HDTV. A very clear separation of the techniques into four parts, related but distinct may be seen being Production, Network Transmission, Delivery to the viewer and Receiver displays. Contrary to the current television systems, differing signal forms will exist in each of these areas of interest.

1. Production uses

In television broadcasting this includes remote and studio pickups, post-production to be followed by conversion to the transmission or release format. There is also conversions with film to consider for both post-production and release. The use of HDTV in graphics production for TV also cannot be ignored and stretches HDTV from TV production into the print world, a very exciting extension of mass communication by TV. In production, TV broadcasting is only one use of the material, albeit important, but what is needed is a high-quality video source for image generation, manipulation and conversion. Such a unique standard now exists in the CCIR proposal attached to Report 801 and while some details need further definition, its widespread use, high quality and balance between current cost/quality and possible future improvement seem to guarantee its wide acceptance. Experience to date in CBC in the use of this system has been very encouraging. All that is needed now is the development of digital recorders and better cameras, both likely to occur very soon.

2. Network Transmission

As with current systems, we will need to find ways to carry HDTV signals between studio centers. This question has been covered previously - it must be a member of the family of transmission modules, but with some signal redundancy removed perhaps, in the coding process to render it more economic. Bit rates of 500 Megabits per second are required or analog bandwidths near 30 MHz.

3. Delivery to the Viewers

The work of NHK in the development of MUSE demonstrates that straightforward techniques, based on sound investigations, can achieve very high displayed quality with a bandwidth reduction of about seven-to-one (7:1). BBC, and others, have proposed novel schemes in which only the video data and control data needed to reconstruct the picture are transmitted, resulting in a further bandwidth reduction in so-called Digitally Assisted TV (DATV). This may open the way to an HDTV broadcast system that will be equally at home in DBS, cable, tape, disc and possibly terrestrial broadcast, based on current channels, i.e. channel compatible. It is unreasonable to attempt to base a new service on the concept of strict compatability with NTSC, as this leads to poor and limited HDTV broadcast quality. Significant and worthwhile improvements to the existing NTSC service can be made, however, using the technology developed for components and HDTV. It seems clear, that the transmission to the viewer, will not be identical to the standard of the display or of the studio, but will be a coded signal from which it is possible to develop by inexpensive and simple processing, a high quality image and several channels of high quality digital sound. We are also sure that material produced, or processed in HDTV will be used for many applications beyond HDTV broadcast, making its convertibility of paramount importance.

4. Display Standards

Rigid standards for the display are not required and the receiver industry will certainly develop a range of products having different levels of performance and price. The receiver of the future will certainly be equipped with a display device, Kinescope, Projection CRT or LCD screen, capable of an aspect ratio of 16:9 and of high brightness. Both the viewers and broadcasters will push for it to display standard TV - NTSC, PAL, SECAM, MAC etc., as well as HDTV, and with connections for tape and disc. Doubtless, the receiver processing will evolve also, so that the improvements to existing systems and HDTV will all eventually get introduced. From the start, we believe that it must be multi-standard and agile so that a truly "world-standard" receiver can be achieved, without an array of converters, adapters and impossible combinations.

Regulatory Aspects

The current broadcasting system, in its broadcast sense, dates from an era of "distribution limited" viewing. We are now in an era of "production limited" viewing, with little limitation on channel availability to the viewers home. This has come about by the strides made in cable, DBS, MDS, and particularly tape and disc. We need to rethink the rationale on which current services are delivered and free the broadcaster to participate in the new reality. It may prove more difficult to move the regulators than to solve the technical questions, but recent studies in Canada, France and England have provided considerable insight into possible directions.

Summing Up

The road ahead for the broadcaster of today is a difficult one, as the three major components of the business, production, distribution and delivery, evolve independantly and are driven by different forces. In addition, the competition for the viewers' time is increasingly fierce as the proliferation of video services continues unabated.

The commercial broadcaster is limited in the availability of capital funding to make the necessary changes, largely by business consideration. The public broadcaster faces these necessary changes at a time when the role and need for their services are under hostile scrutiny in many parts of the world and funding is being squeezed in all areas. All broadcasters will thus be challenged in their quest to maintain a viable broadcast industry, that will provide the services wanted by the viewers.

We have outlined the technical choices and indicated those that we believe are appropriate but the long-term good of the industry requires also evolution in the basic structures of our business. That is the next challenge for us all.

Born in Montreal in 1938, Guy Gougeon graduated from McGill University in 1962 with a degree in electrical engineering. He joined the CBC Engineering Headquarters that same year as an engineer; in 1968, he was promoted to the French Services Div. as assistant divisional engineer and, six years later, as director of engineering. In January 1982, Mr. Gougeon returned to CBC Engineering as vice-president, engineering, the position he now holds.

Cultivating the Wasteland with Technology

Dr. Richard R. Green
Public Broadcasting Service (PBS)
Alexandria, Virginia

Let me begin by asking if you have ever heard of the seersucker theory. It was first published in 1980 by an associate professor at Wharton. It deals with the ability of experts to make accurate forecasts.

The seersucker theory says that, in a given field, the average expert makes a much better forecaster than a layman, but compared to the average student, the expert is only marginally better.

Where the expert's value shows up is not in predicting the future, but in assessing the present. That's where expensive consulting dollars should go. If you have to hire consultants to predict the future, hire cheap ones. They're just as accurate as the expensive ones.

That's why its called the seersucker theory. Because no matter how much evidence exists that there's no such thing as a seer, some sucker can always be found to pay for one.

Having established that my guess is every bit as good as anyone else's, let me now turn to a discussion of the future.

We seem to be back at our game of trying to imagine the future, since the future is not accessible to any other kind of mental inspection. The laws of imagination are not those of logic and reason but of desire and fear. We desire one kind of future, we fear another. Perhaps it will be best to attempt to first understand where we are at the present moment before confronting possibilities, even probable possibilities.

In his recent book, Communications Deregulation-The Unleashing Of America's Communications Industry, Jeremy Tunstall refers to the "twin revolutions of technology and deregulation." It is the combination of running technology on "fast forward" while running deregulation on "fast rewind" that, according to Tunstall, has produced radical change in the communications industry.

In a speech before the International Institute of Communications, Bernard Ostry, chairman of TV Ontario said: "Old boundaries of technology and technical standards, corporate domains, and even national boundaries are increasingly irrelevant. Antitrust constraints are being relaxed; companies are merging, diversifying, and entering into joint ventures to exploit economies of scale and technological expertise. What was once a scholar's vision of global village has turned out to be a businessman's vision of a global marketplace."

In the United States, mass communications from its beginning was privately developed as a vehicle for advertising consumer products. Public broadcasting was an afterthought, a kind of welfare recipient.

In most other industrial democratic societies, early radio and TV service was publicly supported and its content directed at education and the "cultural" development of the population, as well as the entertainment of the masses. Communication technology was a scarce and valuable resource because of the finite broadcast spectrum. Since then the status of public broadcasting has changed radically.

Mr Ostry continues, "Public broadcasting in all industrial democracies is now on the defensive; everywhere lacking the resources, freedom and authority to exploit the new technologies for social ends." In the face of these difficulties, it is often easier to fear the technology of the future than it is to plan and apply the developments which will best serve our future needs.

Of course, fearing change brought about by technological developments is nothing new. I recently came across a letter purportedly sent in 1829 by Martin Van Buren, then the Governor of New York, to President Andrew Jackson:

Dear Mr. President:

The canal system of this country is being threatened by the spread of a new form of transportation known as "railroads". The Federal Government must preserve the canals for the following reasons:

One. If canal boats are supplanted by "railroads" serious unemployment will result. Captains, cooks, drivers, repairmen and lock tenders will be left without means of livelihoods, not to mention the numerous farmers now employed in growing hay for horses.

Two. Boat builders would suffer and towline, whip harness makers would be left destitute.

Three. Canal boats are absolutely essential to the defense of the United States. In the event of the expected trouble with England, the Erie Canal would be the only means by which we could ever move the supplies so vital to waging modern war.

For the above-mentioned reasons, the government should create an Interstate Commerce Commission to protect the American people from the evils of "railroads" and to preserve the canals for posterity.

As you may well know, Mr. President, "railroad" carriages are pulled at the enormous speed of 15 miles per hour by "engines" which, in addition endangering life and limb of passengers, roar and snort their way through the countryside, setting fire to crops, scaring the livestock and frightening women and children. The almighty certainly never intended that people should travel at such breakneck speed.

Signed,

Martin Van Buren, Governor of New York

As a further example, I made the mistake of telling one of my PBS colleagues that I was planning to discuss the future application of technology at this meeting. "You've got to be kidding" he laughed, "nobody in broadcasting wants to hear that we need to spend money on new equipment." After all, most of us think that new technology is just the replacement of one damn nuisance with another. Then he recalled how John Kenneth Galbraith has personalized this idea when he said "the enemy of the market is not ideology, but the engineer." Furthermore he told me, the viewer views programs, not technology and if we're going to spend money it should be on programming, not on equipment. He concluded by saying, "the trouble with you technical people is that you are wonderfully equipped to answer the question how, but get totally confused when asked the question why."

Now you can imagine that his exchange left me feeling a little defensive. My friend had raised a series of valid concerns and what I plan to do today is address the question why as well as how. Why do we need to concern ourselves with the onslaught of new technology?

Why should public television concern itself with the application of new technology? One answer is that the timely application of new technology to the PBS system has served it well in the past and has contributed to its mission and therefore its success.

Public broadcasting has pioneered the technologies which served the producer, the artist and the viewer, and not the profit motive. We must march to a different drum.

I believe that at least part of the credit for the success of PBS can be attributed to the good sense of my predecessors who were able to apply new technologies in an effective way. The innovative approach to station interconnection using a satellite delivery system improved the technical quality and effectiveness of PBS stations. In addition, it opened the door to a series of innovations.

We also pioneered high-quality stereo audio. I believe that our digital audio system, DATE, is still the best television audio delivery system in the country.

Public broadcasting has also helped pioneer and develop the technology which provides closed captioning for the deaf.

If you will forgive me. There is a general perception that all good technical ideas originate in the private sector. Many good ideas have come from the public sector and many significant technical applications have been achieved first by public broadcasting.

Public broadcasting plays a unique role in developing and applying new technology. It is important to recognize that in the rush to embrace free-market principles in broadcasting, the concept of serving the public has been trampled into extinction.

With this in mind, let me characterize what I think the technical future holds in store for us. It is important that future developments in television technology address the public interest as well as the profit motive. This is our specific objective at PBS. We plan to do this in several ways.

First, we intend to foster developments which will find application in instructional and educational television. The use of data channels delivered via educational television signals to augment, complement, or enhance the instructional use of television is of particular interest. Application of compact disc data storage to low-cost instructional television programming, or downloading of program transcripts and supplementary information, for teachers is being studied.

Together with public television station WGBH, in Boston, we have also been experimenting with a new service to provide audio descriptions of the television images to people with impaired vision. A separate audio channel carries descriptive accounts of the pictures to aid in the understanding of the broadcast. Multiple language broadcasts are also on our agenda for the future.

Secondly, we are planning the next generation of our satellite interconnection system. Our present system, in operation since 1978, is showing signs of obsolesence. The new system will become operational in 1992 and will bring with it a new generation of possiblities for service to educational users as well as the possibility of very high quality video for our national program service.

Although it is very early in the planning process, we are looking at several areas of expansion. At present, the PBS interconnection system serves four regional areas: eastern, southern, central and pacific mountain. In the future, we expect that program services will be provided to smaller geographic areas in addition to the national and regional coverage we now provide. For example, states are beginning to recognize the economic advantage in statewide distribution of television by satellite as opposed to terrestrial distribution. Public broadcasters in Kentucky, for example, have led the planning for a statewide educational network using Ku-band distribution. New York, New Hampshire, Ohio, and others are studying similar distribution plans.

The Public Broadcasting Service of the future will probably require more uplinks and more transponders in order to add state service to our present coverage. In addition, the Ku-band offers some advantages for delivery directly to elementary-secondary schools. The increasing availability of easily deployable Ku-band uplinks makes contributions to PBS on Ku-band more likely as well. Therefore, we are studying the possibility of hybrid operation which would integrate the advantages of our C-band plant with future application best served by Ku-band technology.

We are just completing installation of a new technical center in Washington. The structure and organization of this broadcast plant is radically different from our previous center. Our design has emphasized automation and hardware systems which can be upgraded by replacement of software. The plant is served by a software-controlled routing switcher. Each control room operates as a remote terminal sending instructions to the central router. All video tape machines are controlled by a central machine control. We hope that this design will have the advantage of extending the useful life of the hardware while permitting growth in capability through modification and upgrading of the software.

We plan to automate three time-zone-delay feeds providing automatic recording of the eastern time-zone broadcast, and automatic playback for the central, mountain, and pacific time zone programs.

In the area of production, PBS intends to continue efforts to foster an international standard for high definition television. The international exchange of programs in a common video format is a goal of public broadcasting, not only in the United States, but throughout the world. An international HDTV standard is of particular importance to public broadcasting as evidenced by the recent activity in the CCIR. In these meetings, virtually all broadcasters were united in their endorsement of the concept of a single world-wide standard.

Much of the difficulty in achieving a standard can be attributed to private sector concerns and agendas. Concerns with transmission, I believe, can be addressed in an evolutionary way, but the need for a single production standard is immediate and important and cannot wait until we solve the whole puzzle.

High definition television is more important than just a medium of program exchange. This new technology offers some significant advantages to PBS programming. The impact of an opera or ballet is limited by the ability of the present TV system to represent the performance in both picture and sound. The problem is that the details of the costumes and set are lost in present television coverage. One way to solve this problem is to photograph ballet, as we do sports, "up close and personal". In other words, frequent closeups provide the details but fail to represent the interaction among dancers and it is certainly undesirable, if not impossible, to follow the dancers in close-ups during rapid movements.

But with HDTV you can have both. The performance gives the audience a live "I'm there" feeling and the beauty and lavishness of the costumes and sets add immeasurably to the enjoyment of the viewer.

PBS pioneered the distribution of high-quality audio using PCM digital audio transmission and FM simulcast, because the available audio distribution systems were woefully inadequate to bring the full audio performance of a symphony orchestra into the home.

I believe that the present NTSC television system is woefully inadequate to bring the visual performance of a ballet or opera into the home. We need a high-quality video service. HDTV is one answer to the problem.

Whether part of development of an HDTV transmission standard or a separate project, we need to develop a high quality system for delivery of digital audio to the home. The CD revolution has done us in and the consumer is soon going to demand the same level of audio performance from broadcast television.

Finally, given the particular dilemma of public broadcasting in the United States -- that it is a public service technical mission but spartan capitalization -- What is the solution? I believe that new degrees of cooperation are essential with other public broadcasters world wide and joint ventures with our commercial colleagues.

PBS is a participant in the Center for Advanced Television (CATS). You have heard the report from Professor Schreiber on the progress being made at MIT. CATS is a model of the cooperative research programs involving several companies which are needed to support basic R&D initiatives. I would label this activity precompetitive as opposed to noncompetitive, which means that the technical base being addressed is outside the capability of any single company to develop.

I envision that PBS can serve as a buffer in these enterprises, a sort of precompetitive halfway-house. We are in a position to further basic research by taking the effort one step further into its initial application, developing prototype equipment and testing initial applications of broadcast technology.

In addition, PBS will continue its commitment and support of national and international committee activities to further the interests of all broadcasters. We would hope to reduce the chaos which results when technology is on fast-forward and regulation is on fast-rewind. We need technical standards and we strongly support efforts to achieve them.

For example, since no standard was developed for the new 1/2" professional tape format, the process of selection is quite complex. We would like to incorporate that format into our new plant. It does not appear that either 1/2" tape format available can address all our tape needs. It is probably better for us to select digital VTR's.

I personally have invested substantial time in pursuing the international standard for digital television. I hope that the benefits of that standardization effort will accrue to broadcasters and will not lead again to chaos. The lower level of the heirarchy in the 601 standard is still undefined. I hope that we can include in that level provision for both component and NTSC composite recording, which will integrate with the remaining 601 digital standard.

In summary, we will continue our tradition of developing and applying technology which will best serve our producers, artists, audiences, as well as the public interest. We hope to team with business in precompetitive ventures within the broadcast industry and in co-development initiatives in the international community of broadcasters to further these ends.

In a speech given 28 years ago, Edward R. Murrow suggested a quantum upgrade in the quality and realism of television. He concluded that speech by saying:

> "This instrument can teach, it can illuminate, yes and it can even inspire. But it can do so only to the extent that humans are determined to use it to those ends. Otherwise it is merely wires and lights in a box."

I hope that when we look back at the present from the 21st Century, that the technical contributions which we have made were not just to further the "wires and lights in the box" but to give the instrument new vitality, new purpose, and new life!

Richard R. Green joined PBS in October 1984 as senior vice-president of broadcast operations and engineering, where he oversees the PBS broadcast operations and engineering departments, which manage PBS's Technical Center and the PBS satellite interconnection system. Green also supervises the development of PBS technical policies, provision of technical support services to member stations, and research into the use of new technologies for public television.
Before joining PBS, Green took a special assignment to establish the Advanced Television Systems Committee, an industry-supported organization founded to develop voluntary national technical standards for advanced television systems. Before that, he was director of the CBS Advanced Television Technology Laboratory, in Stamford, Conn., from 1980 to 1983.
Most recently, Green was manager of ABC's Video Tape Post Production Department in Hollywood, Calif.; and from 1972 to 1977, he did basic research in laser technology for the Hughes Aircraft Co. in Los Angeles.
Green holds a B.S. degree from Colorado College, an M.S. in physics from the State University of N.Y. at Albany (1964), and a Ph.D. from the University of Washington (1968). A member of Phi Beta Kappa, the American Association for the Advancement of Science, and the Society of Motion Picture and Television Engineers, Green is the author of more than 35 technical papers on topics ranging from TV production to electro-optical and laser research.

Television: The Challenge of the Future

Joseph A. Flaherty,
CBS Engineering and Development Dept.
New York, New York

ABSTRACT

Television is a technology in its youth -- only 50 years
have passed since the introduction of a regular broadcast
service by the BBC. Today, many improvements in broadcast
television services are technically possible. As the 21st
century approaches, implementation of these improvements is
essential if broadcasters are to remain competitive.

The first step toward the future is to begin the orderly
changeover of production facilities to allow the production
of programs of a higher technical quality. To a large
degree, this changeover has begun. No modern television
plant today uses strictly analog composite signals for its
operations. Digital signals and component formats are
increasingly being used to provide new capabilities and
improved performance.

As improvements in production systems continue to be
implemented -- including the use of high definition
television -- superior delivery system must also be
developed. VCRs and cable systems have fewer technical
restrictions than broadcast services in taking advantage of
such developments.

This paper examines the current trends in improved
production systems and their implications for the furture;
and discusses the need to provide higher technical quality
throughout the various delivery channels.

From time immemorial, man has been fascinated with the future — that which not yet is, but is to surely be.

Even Shakespeare's Brutus sighed:

"Oh, that a man might know the end of this day's business, ere it come".

But what of television ere it come?

As of today, broadcasters have so far advanced the art that the world over, more people view television than are literate. The television screen is indeed man's window on his world; yet this technology is a 20th century phenomenon, younger by 20 years than this Society itself. Yet by the dawn of the 21st century, television will offer a diversity of services and a technical quality as different from today's television as the introduction of color was from the early monochrome experiments of Baird and Campbell-Swinton.

The terminal years of this century will magnify the inexorable march toward "The Global Village" and its world economy. Broadcasting, as the vanguard of the "Information Society", will face a host of dramatic changes, and our individual and collective abilities to meet and manage these changes will largely determine broadcasting's character and size, and may indeed mark its very survival.

Technology is today so adaptable to our needs that the technical environment of broadcasting in 1999 will be largely what we choose to make it as we expand, innovate, and try new technologies -- adopting some and rejecting others. But in these last twelve years, two things are certain -- technology will continue to accelerate, producing a bewildering array of equipment -- both for the professional and the consumer; and the consumer will adopt many of these technologies more quickly than ever before.

It took 80 years for the telephone to achieve a significant penetration in the United States. Monochrome television took 20 years, and color 12 years. The VCR will achieve a comparable penetration in 10 years, while ENG swept the country in only 5.

This last quarter century also saw the development of international and domestic communications satellite networks, and these networks have greatly increased the number of wide band communications channels which, in turn, have resulted in a plethora of new program services to the home via cable and pay cable systems.

The broadcaster's monopoly of video channels to the home is gone, and gone forever! The "Television of Abundance" competes directly with the broadcaster for the attention of the viewer.

Change, then, is the operative word. Beware the merchant of the "status quo", for the "technological tranquility" of the past has given way to today's "technological churn."

Ten years ago, at a television symposium in Japan, the world's recording experts examined the prospects for digital video recording and did not envision that future videotape developments would ever produce adequate packing densities or recording heads to record a complete digital video signal -- even in its composite form. Their viewpoint was summarized in the symposium record as follows:

"Since digital PCM video involves bit streams over 100 megabits per second, recording ... cannot be accommodated on an economically viable basis without the need of some form of bit-rate reduction -- a reduction which today results in some level of picture impairment".

Within five years a 4:2:2 component digital videotape recorder was recording a digital bit stream of 240 megabits per second. This machine demonstrated the feasibility of the digital video component signal standard now recommended by the CCIR as Recommendation 601.

Today's videotape technology permits recording of the full bandwidth digital component signal with four digital audio channels on 19mm tape in conformance with the new CCIR digital television videotape recording recommendation.

Yet, lest this achievement be considered "ultimate", last year an experimental digital videotape recorder was demonstrated capable of recording and reproducing a digital high definition television signal at a bit rate of over one gigabit per second. This development represents a ten-fold improvement in videotape technology over that believed possible a decade ago.

Thus, we seem too often prone to underestimate the pace of technological development. That which we believe may never be is soon commonplace, and that which we believe is 20 years hence, is history in five. This was perhaps never so well put as by a man who has truly seen the world in perspective. In September, 1979, astronaut Neil A. Armstrong said:

"Science has not yet mastered prophecy. We predict too much for the next year and yet far too little for the next ten".

What then is the television landscape?

- Production of programs is increasingly becoming a separate business from that of broadcasting or distribution.

- Broadcast distribution is becoming more complex, and, like cable, has replaced its terrestrial networks with satellite distribution.

- CCD cameras are sweeping the consumer and ENG markets, and will find widespread application in the next generation of professional studio and field cameras. Even high definition cameras will soon employ high quality CCD pick-up devices.

- The growth and sophistication of electronic special effects, graphics, and animation continue apace, and their application is increasing proportionally.

- Microprocessors and solid state memories expand their application in virtually all professional equipment and are moving swiftly into consumer electronics.

- Analog component videotape systems are rapidly replacing composite systems in ENG, and newer VTRs employing metal particle tape are challenging the quality of today's one-inch "C" format videotape machines.

- The proliferation of professional videotape formats defy the continued existence of a universal format, and multiple formats will find application throughout production and distribution operations.

- Digital video equipment already in widespread use throughout all television plants will continue to replace its analog counterparts on the way to the all-digital plant.

- Digital videotape machines will soon be widely available and will shorten the competitive life of today's composite analog videotape machines.

- The rapid growth of component video systems in the analog domain will further accelerate the use of component digital equipment including component digital videotape machines designed in conformance with CCIR Recommendation 601.

- Small format (1/2") digital videotape machines will drive a changeover to the digital domain for commercial, promotional, and ENG applications. 19mm videotape cassettes designed for high quality program production will be too large and too expensive for commercial, promotional and ENG use. A new 1/2" digital videotape format is required to service these applications.

- Cinema production, television long-form drama, prime time production, as well as syndicated production and music videos, will make increased use of high definition television to reduce production time and cost without compromise of quality.

- With HDTV production equipment available from nearly 30 companies worldwide, technical studies are shifting to HDTV transmission and distribution systems which will bring the wide screen HDTV experience to the home.

As to prime time television production, the creative community made its choice some 30 years ago. Since then, 80 percent to 85 percent of all prime time programs have been produced in high definition - 35mm film.

However, the core of the problem today is the high and escalating cost of film program production. Increasing at 16 percent per year (figure 1), this cost trend stems from the increased talent, labor, and material costs necessary to assure a more competitive and appealing program product.

Two actions are required. The cost of production must be stabilized and reduced, and programs must be distributed more widely to increase revenue and the return on the production investment.

Today, Hollywood is the world's most successful program supplier. It annually produces over 1,700 hours of original prime time programming for the three commercial networks alone, and yet each one hour prime time series episode costs between $1.2 million and $1.5 million.

Last September, Mr. Larry Gershman, President of the MGM/UA Television Group, said:

"Virtually every program produced for the networks today (is) deficit-financed -- at least $100,000 per episode, sometimes $200,000, sometimes $300,000".

The producer's costs are retired and their profit generated through domestic and international syndication sale. This is how Hollywood continues to be the world's most successful television programmer.

Presently, 63 hours of U.S. produced programs are broadcast each day in the United Kingdom, France, West Germany and Italy. In West Germany, 25 percent of the broadcast schedule is of US origin.

Thus, even a market the size of North America requires international program sales to create competitive programs. It is, therefore, doubtful that any nation can long continue to produce programs solely for domestic or regional broadcast.

It is thus evident that a wider audience must be secured for program distribution. To assist this process, however, it is essential to lower the cost of program production. While electronic production using current television standards can achieve significant financial savings, producers have been, and still are, unwilling to record their programs on present 525 line or 625 line television standards, because the technical quality is limited and indeed suitable only for domestic broadcast.

Thus, any producer seeking a residual, after broadcast market for his program, will not record on any format technically inferior to 35mm film. The 1125 line HDTV system offers this necessary production quality.

High definition electronic production significantly improves the productivity and economy, and reduces the elapsed time of production. Electronic production and post-production of single camera drama offers savings of 15 percent of total production costs.

Single camera production with all the creative flexibility of film will, in the future, be able to serve the producer both on film and on videotape. This production choice will be influenced by the fact that all of the new distribution media -- cable, VCR, DBS, video disc, and of course, broadcasting -- are all electronic. While film will be important for years to come, it will yield a large share of its market to HDTV production. This will be especially true for release through the electronic media, thus avoiding the conversion from film's subtractive color system to television's additive color system, and avoiding the poor motion portrayal artifacts of film's 24 frames per second rate.

The influence of the electronic distribution media is also being felt in cinema production. Figure 2 shows the state of worldwide revenues returned to the Hollywood studios from cinema box offices. They show a slight decline in constant dollars. Income from pay cable distribution makes a modest contribution, but when revenues from the sale of video cassettes is added, it is seen in figure 3 that in the past year they have exceeded cinema revenues and the trend is upward. This trend is developing just in time, for one of the world's largest producer/distributor, MCA, has announced a $50 million write-off for the last quarter of 1986, because of a slump in the domestic syndication market.

In recognition of these important developments, Mr. Jack Valenti, President of the Motion Picture Association of America, said last month:

"One of the most exciting developments that is almost certain to occur within this decade is high definition television (HDTV). The substantially improved resolution related to doubling the number of lines per frame, the increased bandwidth ... etc., offers many new opportunities for our industry. Among these is the increased use of video for the production of feature films. We look forward to the adoption of a single worldwide standard for television program distribution. And in the not too distant future, HDTV coupled with high quality projection systems now in development and large-capacity communication satellites may bring about the development of 'electronic theaters' with significant benefits to distributors and exhibitors."

There is no question that the high quality and the remarkable economies afforded by high definition electronic production are driving a revolution in television and the cinema, and the HDTV videotape can be readily used for distribution through all the electronic distribution media plus that of the cinema itself.

As to distribution, heretofore, broadcasting set the technological pace. Technical quality and broadcasting were synonymous. Cable systems and VCRs were designed to "match" broadcast quality and to be "as good as" the home receiver. Pay cable programmers use broadcast equipment to originate their programs -- to "measure up" as it were. In short, broadcasting and its competitors use the same technology to deliver similar quality.

But all this is changing! New and better technology is becoming available, and the technical quality of services delivered to the home will become an ever increasing factor in audience appeal, and thus in audience size.

As we evaluate the on-rush of new technologies, we must bear in mind that the "standard of service" enjoyed by the viewer today will not be his "level of expectation" tomorrow. I think that most of us would agree that our intuitive measure of picture quality is the cinema -- not television. What will our audience do when they can have wide screen, stereophonic, cinema quality at home? The viewer's "expectation level", not the present "standard of service", will drive our future market.

"Good enough" is no longer "perfect" and may, in fact, become wholly unsatisfactory. Quality is a moving target, and our future judgments must not be based on today's performance, nor on minor improvements thereto. There is neither time nor money to embrace every minor improvement in production and broadcasting technology. We must strive to make major improvements in our service. Technology must be proportioned to the quality and purpose of the program.

Modern multi-channel cable systems are already well positioned to distribute HDTV programming to the home as well as to commercial organizations and to the electronic cinema. In this latter case, the electronic distribution can be encrypted to prevent program piracy. Tests are already underway to explore this new cable technology.

HDTV video disc players are in prototype form, and no fewer than three consumer units were shown this year (figure 4). Likewise, a consumer HDTV VCR with a two-hour playing time has also been demonstrated.

DBS satellites offer an ideal distribution path to the home for HDTV, since only a single DBS channel is now required to carry an HDTV signal.

Thus the terrestrial broadcaster, with the most restricted bandwidth of any wide band mass communications to the home, faces a real challenge in delivering HDTV. Will today's terrestrial broadcaster be relegated to a lesser service when HDTV becomes the dominant home entertainment medium?

American broadcasters don't think so!

Last month, the NAB and MST conducted an extended HDTV test and demonstration in Washington, D.C. Bandwidth compressed MUSE HDTV was transmitted terrestrially over two adjacent UHF channels and via a 12 GHz DBS-type transmitter. Excellent results were obtained with both systems.

The terrestrial transmission of HDTV is technically feasible, and U.S. broadcasters will make it work.

While the UHF solution would involve a major reorganization of the frequency band, the use of a portion of the 500 MHz in the 12 GHz DBS band for terrestial transmission is possible and was approved by the ITU at the WARC '73. Naturally, the FCC would have to reallocate a portion of this unused band for terrestrial use, but I believe that this is possible.

With technology such as the NHK MUSE system or the BBC DATS system, an HDTV signal can be transmitted over a single 24 MHz DBS channel – a far cry from "it can't be done" which was heard scarcely five years ago. As to terrestrial propagation of the 12 GHz signal, the first definitive test of 12 GHz television transmission was made here in San Francisco by the Westinghouse station, KPIX, and CBS in 1982. The results were reported to the FCC, and further tests with high power transmitters could extend the work already begun here.

The final transmission signal format has not yet been chosen, as that is the task for the next two years. The work is underway at the ATSC, in Japan, and in Europe. Several proposals have been made and more will come. Some may offer compatibility and some will not. I would only point out that most great advances were not compatible with their predecessors. The automobile was not an extension of the horse and buggy, nor was the steamship compatible with sailing vessels. FM broadcasting was totally incompatible with AM radio, requiring an entirely new receiver. Today's revolutionary compact disc is incompatible with the phonograph record and audio cassette, yet compact disc players and the digital discs are sweeping the market before our very eyes.

Naturally, there is no evil in compatibility, but today's sophisticated consumer is undaunted by totally new technologies as long as they are sufficiently better to prompt him to buy. I believe wide screen HDTV with digital stereophonic sound, able to be displayed on a large screen is such a technology. It will prosper with or without compatibility!

As to standards in all fields of television, the exploding technologies are not on trial. They are being created at an unprecedented rate and exist before us. Rather it is the standards organizations worldwide that are on trial.

If they cannot reorganize themselves to cope with the
stampede of new technologies they will simply be trampled.

The broadcasters no longer have a benovolent FCC to
set and enforce technical standards, for the FCC has a
marketplace standards philosophy. Rather we must set our
own technical standards in organizations such as the
ATSC, the CCIR, the SMPTE, etc., or not have them at all.

Herein, the users bear an especially awesome
responsibility, since it is the users who pay the bills
and it is not always in a vendor's best interest to have
a standard — especially if he has committed large sums of
money to a product not endorsed by the standard, or if,
in desperation, he needs to protect a shrinking market.

In any case, our standards record is not
umblemished. We failed to produce an AM stereo standard,
a teletext standard, a 1/2" videotape standard, and even
our beloved one-inch "C" format standard is only a "De
Facto" standard.

If our past is marred, what of our future?

While Sallust encouraged Cesar:

"Experience has shown .   .   . that every man is
the architect of his own future."

Coleridge reminded us in 1831 that:

"If men could learn from history, what lessons it
might teach us! But, passion and politics blind our
eyes, and the light which experience gives is a
lantern on the stern, which shines only on the waves
behind us".

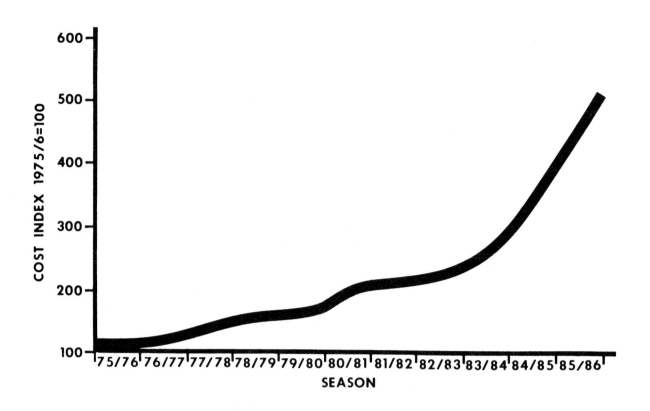

Fig. 1. Network prime time series production costs.

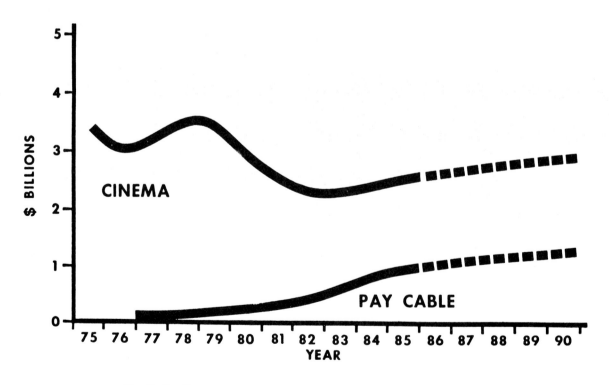

Fig. 2. Studio revenues from cinemas and pay cable, worldwide. (1985 $)

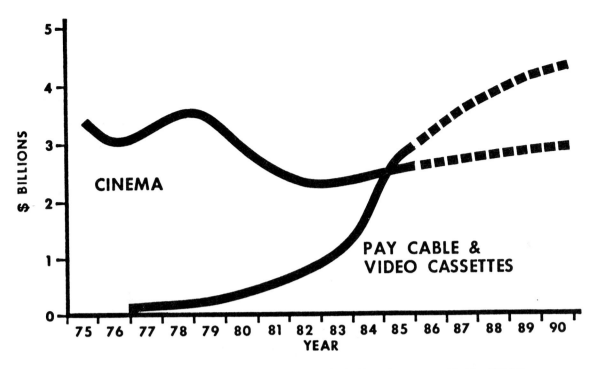

Fig. 3. Studio revenues from cinemas, pay cable, and video cassettes, worldwide. (1985 $)

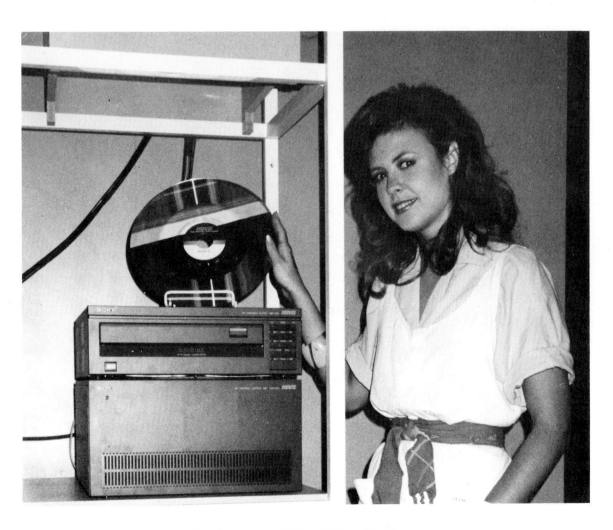

Fig. 4. Prototype high definition videodisc player.

Joseph Flaherty received his degree in physics from the University of Rockhurst, Kansas City, Mo., in 1952, and began his television career at WDAF-TV. After serving with the U.S. Army Signal Corps. in 1955, he joined NBC-TV in New York as a television engineer.

In 1957, he joined CBS as a television design engineer. In 1959, he became the network's director of technical facilities planning, and in 1957, he was promoted to general manager, and subsequently appointed vice-president and general manager, engineering and development.

Flaherty is a Fellow of the SMPTE, and has served as SMPTE Vice-President for Television Affairs and as SMPTE Executive Vice-President. He is a Fellow of the Institution of Electrical Engineers (IEE), U.K.; a Fellow of the Royal Television Society (RTS), U.K., a member of the Society Fernseh-und-Kinotechnischen Gesellschaft (FKTG); F.R. Germany; a member of the Societe des Electriciens, des Electroniciens et des Radiolectriciens (SEE), France; and an Honorary Member of the Institute of Television Engineers of Japan.

Flaherty serves on the Executive Committee of the Montreux International Television Symposium, the Excutive Committee of the Advanced Television Systems Committee (ATSC), and the USIA Advisory Committee on TV Telecommunications.

In 1969, he was the recipient of an Emmy Award Citation for the CBS Minicam Color Camera, and was the 1974 recipient of the David Sarnoff Gold Medal. He received, for CBS, the 1975 Technical Emmy Award for ENG, and in 1979 he received the Montreux Achievement Gold Medal for the development of the concept and the operational implementation of ENG. In 1983, Flaherty was the recipient of the NAB Engineering Award; in 1984 he received the SMPTE Progress Medal Award. In 1985 Flaherty was nominated to the rank of Chevalier d l'Ordre National de la Legion d'Honneur by President Mitterand of the Republic of France. In September 1986 he received, on behalf of CBS, the Emmy Award for the electronic editing system for programs produced on film.

He is a frequent lecturer on television technology and has published many technical articles on various aspects of television broadcasting.

New Technology and the Broadcaster

Max Berry and Robert Thomas
Capital Cities/ABC, Inc.
New York, New York

ABSTRACT

In the formative years of television broadcasting, new developments in technology were immediately and enthusiastically embraced by engineers and management alike in their drive to improve picture quality, extend services and add new production values to television programs. Emphasis is now increasingly being directed toward the impact of technology upon spectrum conservation, viewer perceptions and the vital matter of the profitability of broadcast companies themselves. New developments must be analyzed in these terms and realistic priorities established to identify those projects which should be implemented immediately, relative to those that might be postponed or abandoned entirely. Equipment designers and manufacturers are challenged to produce hardware that translates the promise of new technology into products that will justify capital investment by a return in increased reliability and reduced operating and maintenance costs.

+ + + +

In the past forty years there have been rapid developments in the parallel fields of general electronics and broadcast engineering, whereby advancements in electronics technology have found a ready application in functional broadcast equipment. For example, vacuum tubes were rapidly supplanted by a succession of new solid state devices in the progression from discrete transistors through integrated circuits, LSI and VLSI. Advances in circuit realization enabled bulky hand-wired assemblies to be replaced by compact automated assemblies incorporating rigid and flexible multilayer etched circuits, surface-mount components and miniature connectors. As the science of filter synthesis developed, television equipment designers were quick to apply these new techniques, as was the case with the technology of low noise amplifiers, high performance servomechanisms, and efficient magnetic devices.

In each of the examples cited above, there was an immediately obvious advantage in adapting the newly available electronics technology to broadcast equipment in terms of reliability, performance, size and weight, capital investment and a well defined financial return from reduced operating costs. In short, the consequences of incorporating new technology were obviously beneficial from the viewpoints of both engineers and management.

Innovation in broadcasting was not confined to the application of advances in general electronics to specialized broadcast equipment. Rapid strides were also made in what may be termed "pure broadcast technology", that is, those developments that were conceived and applied specifically for the benefit of television broadcasting. The most notable of these is the system for compatible color television, developed by the industry under the leadership of RCA, and presently employed in most countries under standards of the NTSC and their derivatives.

Other major broadcast-specific developments include elaborate systems for signal switching and distribution, graphics, digital effects and timebase synchronization; camera pickup devices; video tape recording; electronic news gathering; satellite program distribution; special test and monitoring equipment; and operating procedures that ensure consistent quality in the utilization of all this specialized technology. Again, these new developments were enthusiastically embraced by engineers and their management because of the indisputable advantages they conferred beyond the status quo. As a result of these fundamental advances, broadcast operations and engineering have grown to a mature and profitable business of television program delivery.

Not all new developments have shared the success of the innovations described above, even though they might have possessed refined principles of engineering, or furnished unique broadcast services. They withered because they failed to fulfill essential criteria of public acceptance and viability in a highly competitive environment.

A prime example of inappropriate technology is Teletext, a derivative of a dream of the late 1930s, whereby newspapers were to be delivered to the home by facsimile. Teletext was destined to fail in the United States for two principal reasons: First, the public had faster access to better information without having to devote their undivided attention to a screen or become entangled with a complex menu-driven system that even many engineers found perplexing. Second, and even more decisive from the standpoint of management, was the inadvisability of furnishing a secondary broadcast service that might encourage viewers to tune away from the main program channel, especially during commercial messages! In recognizing these innate deficiencies, Julius Barnathan, President, Broadcast Operations and Engineering at Capital Cities/ABC, avoided an enormous capital investment by ABC in a technology that was utterly inappropriate for the U.S. market. For engineers, the legacy of Teletext is, hopefully, a sense of respect for the business facets of new technology.

Many current developments which, to many engineers appear to hold great promise, also suffer from intrinsic frailties when viewed in the cold light of viewer needs and perceptions, integration with existing systems, operational complexities and the ever present matter of economics. There was a time when some of this new technology might have slipped unnoticed into day-to-day operations, but recent developments on many fronts have placed all proposals for the expenditure of funds under elaborate evaluation at all levels within broadcasting corporations.

The technical infrastructure of broadcasting has to be viewed as a business enterprise responsive to the benefits of rigorous financial administration, in which cost control is a prime element. Capital investment is encouraged if it is essential for maintaining existing operating capabilities, reducing current operating costs, or increasing revenue by attracting more viewers or sponsors; investment for other purposes must be justified by the unique circumstances that require the funding. These potent criteria are the "Facts of Life" for the U.S. broadcast engineer, and they directly affect equipment suppliers and common carriers as well. Three basic processes are involved in the application of the criteria to new technology:

First, a distinction must be made between that new technology which is essential to the survival of broadcasting, and that which could be abandoned in the interest of practicality, or at least deferred until a more favorable business climate exists. For example, the absolute reliability necessary for airing commercials could dictate immediate replacement of an aged video cart machine, whereas, conversion from existing operational U-matic ENG equipment to a newer format might be postponed with relatively little effect upon daily operations.

Second, a critical assessment must be made of projects that are, in essence, very complex and expensive approaches for solving problems of questionable importance. Broadcast companies cannot afford to dissipate their engineering resources on the solution of operating problems that can be ignored entirely, circumvented, or at least made tolerable, by imaginative use of existing facilities. As will be shown later, component analog video systems fall within this category.

Third, broadcasters must re-assess those large scale development programs, such as high definition television, which are dedicated to the promotion of new television systems that are basically inappropriate for the broadcast medium. The burden of support for such developments should be upon the principal beneficiaries, relieving the broadcast sector of involvement with technologies that are not in concert with the fundamental precepts of broadcasting.

Certainly the two most important assets possessed by the television industry in the United States are the NTSC composite color signal and the existing technical facilities. NTSC color might well be characterized as the most elegant system ever devised by engineers, for in one complex electrical signal, in one channel, are all the attributes of a picture in full color. Furthermore, as with any true classic, the more intimately an association is developed with the NTSC signal, the more readily its full potential is realized and ultimately applied in practice.

The NTSC signal is handled routinely in every broadcast system in the country - systems that are in place, paid for, and well understood by legions of operators and maintenance personnel. These systems range in complexity from single room operations to elaborate network facilities, comprised of dozens of studios, hundreds of VTRs, and a myriad of graphics and digital effects devices, all tied together through miles of coax cable to a central control and a routing switcher typically having 200 inputs and 200 outputs. These existing systems, regardless of size, all have one thing in common: NTSC video on one coaxial cable.

The two primary elements of television broadcasting - a powerful signal configuration and an existing elaborate technical production system - must continue to be the foundation of commercial broadcasting studios if profitability is to be maximized in the difficult years ahead. There is no denying that engineers can identify certain impairments in picture quality that arise in current programming, either due to the NTSC signal alone, or in combination with processing and distribution techniques. But legitimate questions arise with regard to how serious the problems really are, how they might be eliminated, and what steps might be taken to accommodate them by modifying present operating procedures.

There have been two approaches to impairment problems, although the two are intended to solve somewhat different aspects of the problems. One approach has been to simply escape problems altogether by adopting a completely different signal system utilizing analog components. Another approach has been to examine the basic properties of the NTSC signal and apply great ingenuity in system design to eliminate, or at least ameliorate many observed impairments. A third approach, seldom discussed in engineering circles, also exists. It is to find out exactly what subjective impact these so-called impairments have upon the viewer population, before abandoning a system already known to work well. This principle applies equally to proposals for improvements in NTSC; the cost/benefit tradeoffs must be established before an intelligent decision can be made as to whether or not demonstrated picture improvements are sufficiently justified to warrant the expense of the license fees, modification of existing equipment, and purchase of new equipment required to incorporate required changes.

To date, the industry has plunged ahead as in the past, when there were unlimited funds to support radical system changes. But conditions are different now and many engineers, in response to policies emphasized by corporate management, are questioning the practical value of changes in technology before accepting them. Even when improvements are indisputably justified, there are considerations about the possibility of deferring them to a more favorable business climate. All of this may be summarized by stating that a rational justification is required before the existing system of signal processing and distribution is changed or abandoned.

Assuming justification for improvement in picture quality is indeed eventually forthcoming, the task then becomes one of determining the best way of accomplishing changes that are appreciated by viewers and most cost effective for broadcasters. An approach that has a substantial following at present is use of component analog video (CAV) to varying degrees within the television plant. Reasons for using this technique include the possibility of achieving better chroma key effects, reduction of impairments in edit facilities based on component video recorders, and as an initial step in the evolution of component digital studio systems.

All of these goals have merit, but there are alternatives worthy of new consideration in present circumstances. Improved chroma keys, at least in broadcast post production as distinct from the production house situation, can be realized by modifying operating practices; development of improved decoders and encoders for component VTRs, a process already well advanced, virtually eliminates the need for component edit systems; and component digital systems, once thought inevitable, might be circumvented as the development of composite digital video recorders progresses, enabling the one-wire NTSC system to be maintained in tact in most places in the plant. These suggested alternatives might not produce the **best** results, but they will provide **more than adequate** results which will also meet essential business criteria of modern broadcast management.

An appropriate application of CAV might be for interfacing graphics and animation devices, many of which are component-based and frequently utilized in re-entrant system configurations. That application might best be approached as an "island" concept, in which an isolated graphics/animation facility processes incoming composite video through high quality decoders, and outputs through similar quality NTSC encoders. But even such an apparently innocuous use of CAV propagates complexity throughout the well established composite TV plant, as evidenced by recent proposals to universally eliminate setup throughout the entire industry to cope with a requirement unique to some component systems.

Other proposals have been made for improvements in picture quality while simultaneously preserving the composite NTSC signal and these are certainly worthy of close scrutiny. Specifically, the work in the prefiltering of encoded video by Faroudja[1] might hold promise for practicable quality improvements at the viewer level. This kind of innovative development should be given preferential consideration as an orderly step toward upgrading broadcast video without inflicting expensive radical system modifications upon existing systems, for it is an approach much more in tune with current business realities.

There is much activity in the field of high definition television (HDTV) at this time. Originally conceived as a production standard and now advocated for broadcast, the 1125-line 60-field system that is presently dominant is questionable, both economically and socially, as a broadcast medium. RF spectrum considerations aside, there does not appear to be an economically justifiable rationale for involvement of U.S. broadcasters with such an extravagant project.

However, a considerable amount of research is being conducted on improvement of television pictures while preserving the assets of the NTSC signal. Approaches presently being explored are enhancement of basic video quality, modification of vertical scanning, and optimization of display parameters, such as aspect ratio, for heighten viewer impact. Although they will not produce the quantum improvements offered by HDTV, neither will they inflict wrenching costs upon broadcasters or the public. The imminent introduction of video frame stores and digital processing techniques in receivers will also result in the enhancement of the received signal and its display, to impart a significant improvement in subjective picture quality. Many of these improvements will reside entirely outside the broadcast domain or will exist in receivers, and so will be financed solely by those willing to pay for improved picture quality.

On the broadcast side, a considerable amount of work has been reported by Fukinuki[2,3] which is directed toward an improvement in horizontal resolution of about fifty percent while maintaining full NTSC compatibility. It is this sort of finesse in the realization of the classic potential of the NTSC signal that was referred to earlier, and it is this sort of development that should be seized by broadcast engineers for advancing picture quality without resorting to complex and expensive schemes to overturn established facilities without regard for the practical consequences.

We have now seen that there are already underway at least two developments in broadcast systems which possess, at once, the potential for significant improvement in the quality of program delivery, and the economical integration with existing facilities. Possibilities also exist for further picture quality improvements by the application of adaptive digital signal processing, both at the originating and receiving ends of the system.

Although improvements in picture quality are important, they might not have the highest priority among broadcasters at this time. A far more crucial element at present is the potential for improved profitability from a reduction of operating and maintenance costs. Recent inroads by the new 1/2-inch cassette-based video recording formats have resulted only in part from improvements in picture quality relative to older U-matic equipment. To many broadcast companies, an even greater incentive for adoption of this equipment is the prospect of the reduced operating costs it portends. For electronic news gathering in the field, deployment of one-piece camera/recorders can be translated to a reduction in crew size. And for studio applications, cassette-based VTRs with high performance signal handling capabilities will supplant reel-to-reel machines in selected applications, effecting a reduction in manning requirements; a further reduction in staffing will be achieved by automated multi-play machines.

The degree to which the new formats are applied to daily operations will depend on the type of broadcast operation, its market situation, and policies of Operations management. One network has announced plans to utilize a 1/2-inch format for all applications currently assigned to 3/4 and 1-inch VTRs. Another has indicated the intention to utilize a similar format primarily for commercials and integration of taped segments into news programs. Small broadcasters may find it advantageous to fully convert to a 1/2-inch system for its tape operations. In other cases a more conservative approach will be taken whereby 1/2-inch equipment will be integrated with existing 1-inch machines for optimum utilization of equipment and operating staff in each application. The point is, though, that justification for conversion to 1/2-inch is only partly because of its performance capability; the main reason is reduced operating costs.

Cost reductions are anticipated as digital equipment continues its penetrateon of the broadcast field. Operating costs might respond favorably to digital technology due to greater reliability and the virtual absence of adjustments on digital equipment. Maintenance costs resulting from direct labor and down time should be much lower than is presently experienced with analog equipment due to the inherent reliability of digital circuitry and the capability for inclusion of service diagnostics. But all of these statements are qualified by the terms "anticipated," "might," and "should."

The presumed advantages of digital equipment have become accepted as fact because they seem plausible and have been repeated so often. At the moment, however, we do not really know that the cost of conversion to new digital products will be offset by lower operating and maintenance costs. At the present time, our experience does not indicate that digital is synonymous with reliable. Therefore, broadcasters will look to researchers, product designers, and manufacturers to provide equipment that can be proven to reduce costs. Without a factual assurance of cost benefits, broadcast engineers will not have a sound basis upon which to recommend capital investment in new hardware, regardless of anticipated benefits.

Special note should be made here of the emergence of composite digital video recorders. Combining attributes of digital technology and the composite NTSC signal in a cassette-based format, these economical machines could well have as great an impact upon present recording methods as the C-format had upon quad machines. Additionally, composite digital VTRs can be introduced gradually into existing plants as direct replacements for analog production-quality recorders, with no changes to video distribution systems. But before a commitment is made, the cost benefit must be established with certainty.

These are examples of a thrust from technology that should provide corporate management with a compelling reason for investment of capital in modern technical facilities and, incidentally, for justifying the presence of an engineering department sensitive to the total needs of the company.

Many other avenues are also open for cost savings. For instance network program distribution, presently handled almost exclusively by company-owned satellite systems and leased common carrier facilities, could soon realize immense cost reductions as nationwide optical fiber digital transmission systems become commonplace. ABC is currently investigating this new transmission medium in a cooperative effort with AT&T, utilizing T-45 facilities between New York and Washington, D.C. Evaluation with on-air programming will be conducted for a one year period, following which a determination will be made as to the future utilization of this type of facility. At present, optimism is high on the part of telecommunications engineers that digital fiber optic transmission will yield substantial cost savings.

Another possibility for reduction of operating costs lies in the design of a system for automatic control of a studio camera mount which would continuously follow shoulder movements of a news anchor without operator intervention. Successful development of such a device would derive significant

208

cost savings at all levels of news programming, from small stations to networks, without imposing unnatural restrictions on the anchor or sacrificing production flexibility. It is another example of a development worthy of capital investment because of the potential savings it will generate.

The changes engineers are experiencing today are a severe test of their professional adaptability. In the months and years immediately ahead, management will place increasing emphasis upon developments that translate directly, and preferably immediately, into improved profitability. In such a milieu there will be no place for projects that are intellectually stimulating but demand wholesale revision of established systems; for projects that produce fantastic results in the opinion of engineers but mean nothing to viewers; or for projects that are simply too vast in relation to industry resources.

Quite the opposite of being discouraging, the new realities exist as a challenge to the flexibility and ingenuity of the entire community of broadcast engineers. By successfully meeting that challenge with appropriate technology, engineers will play a major role in rising above our present difficulties and assure their place in a healthier broadcasting business.

REFERENCES

1. Y. F. Faroudja, <u>Optimizing NTSC to RGB Performance</u>,
   (Private Publication, Faroudja Laboratories, Sunnyvale, CA.).

2. T. Fukinuki, Y. Hirano and H. Yoshigi, "Experiments on Proposed Extended-Definition TV with Full NTSC Compatibility," SMPTE J., 93: 923-929, Oct. 1984.

3. <u>Fully Compatible EDTV (Extended Definition Television)</u>
   (Private Publication, Central Research Labs, Hitachi, Ltd., Tokyo, Japan, 1986).

Max Berry is an electrical engineer who has a background in equipment and systems design, marketing and management and has been with ABC for 21 years. He joined ABC in 1965 as manager of equipment planning, was named manager of audio/video systems engineering in 1968, and promoted to director of audio/video systems engineering in 1976. He was appointed vice-president, engineering, B.O.& E., in 1985. During the period from 1970-1987, all of ABC's New York, Washington, London, and San Francisco, technical facilities were either built or rebuilt under his management and control. Mr. Berry was an engineering leader in the Astro Electronics Div. of RCA from 1961-1965, during which time television cameras were designed and built for use in unmanned satellites. Prior to that, he was a marketing manager in the Commercial Broadcast Equipment Dept. of RCA, Camden, N.J. from 1955-1960. This followed a period of four years in the NBC Development Lab. where he worked on scrambled TV systems and built the first color film chain used on the air in the U.S. Mr. Berry has a Bachelor's degree in electrical engineering from Cooper Union and an M.E.E. from New York University.

Robert Thomas graduated from Drexel University with a B.S. in E.E. in 1950. He joined RCA Broadcast Systems Div. in 1951 and was engaged in the design of numerous video distribution and switching products until 1960. He then transferred to the video recording section, specializing in the design of video, FM, and audio signal processing systems. He joined ABC as senior equipment planning engineer in 1982, with responsibility principally for video recording equipment. In December of 1985, he was made director of the newly created Technology Planning Dept. Mr. Thomas is a member of IEEE, RTS, and SMPTE.

New Frontiers, the Next 15 Years

Michael J. Sherlock
National Broadcasting Co.
New York, New York

The paper looks at progress in the past 15 years, examines the likely changes
in the next 15 years, explores the processes of making television, in the
industry and at NBC. Quality improvements and cost reduction as prime
motivators are discussed. New camera and recording techniques, software tools
and automation are included. New facilities of the future at NBC are
discussed, as is the contribution of standards to the industry. A standard
for a 1/2" digital component recorder is proposed.

Good afternoon. As we've heard this afternoon and over the past two days,
television in the late 1980's is developing in many directions
simultaneously. But much of what had to be said has already been said by
those who have spoken earlier and I don't want to repeat their presentations.
I would like to take my shot at discussing the trends going into the 21st
Century - particularly as they fit into the practical reality of today's
broadcasters. Obviously, before we attempt to examine the future, we should
take a short backward glance.

15 YEARS AGO

Going back only 15 years: we were using 2" VTR's; video editing was done by
cutting tape; and the early CMX systems were just being installed. Electronic
graphics was in its infancy. Slow motion was done on disk. Commercials were
delivered reel-to-reel. Distribution was terrestrial. The home VCR was not
on the market. Audio was monaural and 5kHz bandwidth; and NTSC was truly
Never Twice the Same Color since no two picture monitors or receivers
displayed the same hue. The methods of bringing programming into the home
were limited and the majority of the viewing audience appeared to have no
greater expectation.

AUDIENCE EXPECTATION

The steady progress in television technology during the recent years has made
it clear that programming delivered to the home in the near future will be
dramatically improved in both audio and video quality. As the viewing
audience's appetite expands and they become more aware of the choices
available to them, they will demand affordable, improved services. They will
come to expect high-quality, multi-channel digital sound and the removal of
distracting video artifacts, like flicker, cross-color and cross-luminance.
And, because sales promotion is what it is, they will come to look forward to
the home delivery of widescreen, enhanced television or even High Definition
Television. The last two days have shown that the timing on this is uncertain
and there are many unanswered questions about standards, compatibility and the
delivery systems...but we know it is coming and, by our standards (to coin a
phrase), it is coming fast.

Of course, an important issue, especially when it comes to the audience, is progressing to an improved system without making 150 million television sets obsolete! Compatible improvements could be achieved by improving signal processing in the television plant. The Engineering community has made several good proposals in this regard. They include: use of down conversion, application of additional spectrum sharing techniques, digitally-assisted picture enhancement and smart receivers. The maximum improvement from these changes would be seen in these new receivers specifically designed to take advantage of the changed signal, but old receivers should see significant improvements as well. Another welcome improvement for the viewer will be the replacement of present bulky picture tubes by the development of new flat panel displays which will be hung on the wall.

Looking beyond compatible improvements which can be transmitted over one standard television channel, some have suggested the possibility of using two channels to broadcast a closer approximation to high-definition television. Alternatively, high-definition television could be introduced as an imcompatible new service, delivered to the home via very wideband DBS or cable channels and by a new generation of high-definition television disk and tape machines. Freed from the constraints of compatibility, such a system would probably be technically superior to compatible systems. However, without compatibility, such a service might not be commercially feasible because of the usual "chicken & egg" problem encountered in the introduction of any new program service.

Clearly, the technical problems of HDTV can be solved, but a practical business plan, matching audience expectation to his ability to pay has not yet been devised. This can't be dismissed when evaluating the technology because, as good engineers, we are not interested only in technology but also in its application to the business enterprise. I have yet to hear or see a truly practical plan for implementing high-definition television in which I and other broadcasters can participate.

On the other hand, as I've mentioned, the possibility exists that a totally incompatible delivery system, via tape machine or video disc, will be introduced in the United States within 2, 3 or 4 years. Such a system would be in direct competition with free broadcasting and pay cable as we know it today. Left alone, it could create a de facto delivery "standard" without any regard for the procedures we have established to protect against such events. I would suggest that we as television engineers may want to make sure that the change to a widescreen, enhanced service is controlled to be evolutionary and does not occur in a way that disregards our current audience, service and investments.

THE PROCESS

So how are we going to get there? The television image facility can be viewed as a factory. Factories, simplistically, are facilities which take in raw materials and energy and produce finished goods, using machines and people as processing devices. This is accomplished, according to some preconceived plan and schedule, by controlling mechanisms. Raw materials proceed through several different subprocesses, eventually converging to be combined into a finished product.

The raw materials in our factories consist of video and audio program material in different states of development. Certainly, a live feed from a plane crash site is quite different from a feature film packaged in tape form. These raw materials are processed through various subsystems, with eventual play to air. Processing can consist of a simple log sheet entry and a library shelf storage location or many hours of editing, cleaning and eventual integration with other processed materials.

The television factory of the past was conceived, planned, built and operated around some very severe limitations. The video signal required special processing and monitoring. Video equipment such as tape machines, processors and switchers were large, power-hungry, limited in capability and often unreliable. Audio signals faced the same problems and restrictions, but shared no commonality with video equipment itself. Intermachine and interfacility communication consisted of people and telephones with the drawbacks of slow speed, limited capability and susceptibility to error. Television producers and directors were limited by the equipment and the facility.

To a great extent, the tape machine and post-production editing provided some relief to many of these facility limitations. With the advent of solid state devices, equipment has become more reliable and provides greater capability at less cost. Micro-processors and the growth of software development have opened up the arena of intermachine communications and control. The reduction in size of equipment provides new opportunities in news gathering and sports coverage. The pent-up demands of the producers and directors are now being met and, in fact, technology now challenges the producer to be innovative and creative.

Satellite communications have removed many of the limitations of transmission between facilities worldwide. However, the current television factory remains a direct descendent of television factories of the past. It is now necessary to re-examine the role the large television facility plays in the context of its function as a factory and its ability to produce enhanced services. Of course, factors such as a more competitive environment and costs of operation become paramount.

Currently a great deal of thought, research and implementation is occurring in the field of factory systems. Concepts such as CAD/CAM, robotics, computer assisted scheduling, just-in-time manufacturing and artificial intelligence are coming into increasing use. A greater degree of integration, from concept through final shipment of product, increases efficiency and produces greater reliability. We are an information processing factory, where information is transformed and transported to the loading dock of our satellite feed and the technology used for information processing is directly applicable to our business.

GATHERING

Let us now look at that factory over the next 15 years - how we gather the raw materials; what constraints our future products will place on us; and what tools might be available to meet the demands.

High Line Rate Cameras

In gathering the raw images, we must consider that the camera used for high definition or any enhanced television places more severe constraints on the resolution of the optics, pick-up devices and front-end performance. We are going to be demanding new methods of contour correction, dynamic lens error compensation and other camera-related technologies...and all must be automated. Ideally, the CCD imager, with its low lag, will be the pick-up device for enhanced image cameras.

Cameras and Robotics

Getting back to the practical, camera robotics will be applied to the reduction of labor costs. In the future, higher resolution control circuitry of 20 bits or more will permit very precise direction of camera movement. Existing hardware used for stationary studio applications will be refined to permit longer focal length lenses with a high degree of repeatability and accuracy. Sports pickups that normally require a day or longer for setup and strike could be covered by permanent installations using this technology. Not only would savings arise from reduced labor costs, but hostile or inaccessible locations would also be better covered by remote camera systems.

Audio

The word in television audio over the next 15 years is increasingly going to be "digital." The digital trend is already underway with both 1" and the new M-II video tape formats offering the option of 16-bit digital audio recording to provide complete audio transparency, total editing flexibility, multi-generational audio dubbing without degradation and the absence of time delay or phase errors between stereo channels. It is our hope that early in the 15-year period the cost differential for digital audio systems will become insignificant and cabling requirements and monitoring practices will be developed to facilitate implementation.

FUTURE TRENDS IN RECORDING

Magnetic

In video recording, higher recording densities or shorter recorded wavelengths are made possible by utilization of new types of media. Applications to television recording are now overlapping with other applications, such as data recording for instrumentation and computers. The evolution of new magnetic recording techniques for television will be influenced by information processing applications more than ever before.

Optical

The high rate of information retrieval possible from optical disk systems and the potential for an unlimited lifetime make them attractive as a viable television recording technique.

Solid State

Ultimately, a recording system which has infinite storage lifetime and no moving parts should be a part of our future wish list for obvious reasons. Solid state storage will become more competitive, both operationally and economically. My News Department is always asking me for a tape machine with no pre-roll. This form of storage will give it to us.

1/2" Digital Recorders

But, let's slow down a bit and talk about video tape. The next broad application format - or "universal" format - will be required in the not too distant future to complement the super high performance D-1 format VTR. This digital format - to be usable in many applications - should be small - i.e. no greater than 1/2 inch and it should be component. I might also add...it must be inexpensive.

The important issue here is that whatever recorders are developed should be to a common format. Let's not fall into the trap Joe Roizen thinks we're in. As Dick Green & Joe Flaherty said, there is a need to standardize in this area and I urge the SMPTE to start it now or else the pressure of the current market, some of which NBC created, will continue to confuse us all. Let me state that again and more clearly. We need a standard for a half inch, digital component tape format. The SMPTE should take up the matter now!

MOVING THE MESSAGE

The video switching, routing, timing and control systems of the future will be configured in such a way that many of the individual pieces of equipment that today are dedicated to one studio or edit room will be assigned by software control into any form or forms of reconfigurable video installation. I see every piece of equipment appearing on a huge computer-controlled jack-field which has no timing constraints. This will enable broadcasters to buy fewer pieces and use each more. Production switchers will also be integrated into this scheme. This reconfigurable video system leaves only the special effects functions as isolated building blocks.

But, before the beginning of the 21st Century, video effects devices will become non-specific, i.e. their function will be software defined and only limited by computer power. Mass storage devices will form many of the building blocks of the reconfigurable factory.

The messages will be moved within an all-digital component studio. This will include digital tape, advanced digital distribution and switching, improved computer control and digital special effects. The result will be better television pictures delivered to the home, even if it is encoded into NTSC for transmission.

In many ways, the above trends are somewhat analogous to the evolution which spawned personal computers starting from the dedicated, first-generation Nixie tube type calculators. There is a strong family relationship between the storage and transmission of data and the storage and transmission of video. I see this daily when I compare the trends in technology under consideration in my MIS facility and the trends in technology under consideration in my engineering functions. The data bus of a PC can also be likened to television's future routing and distribution systems.

Productivity Enhancements

Recent developments in robotics will make possible sophisticated recording automation systems. Automatic alignment has already been applied to studio cameras and has recently appeared in studio picture monitors. Audio console and video production switcher automation has been gradually gathering momentum over the past few years. Mass video storage systems which will permit recording of minutes of video with frame-by-frame random access promise to revolutionize post-production. Synthetic scene generation is under intensive study. These will increase the efficiency and flexibility of production and post-production systems, permitting innovation in editing techniques.

CONFIGURING THE FACTORY

Well, how does this all convert to real life? As we announced last year at this meeting, NBC is planning for an all-new technical facility for its New York operation. Current plans call for completion in the 1994-1996 time frame. Our challenge will be to achieve the proper mix of flexibility, automation and technology at a reasonable cost to serve our need for the next three to four decades. The need for flexibility to accommodate ever-changing technology will lead us to explore reconfigurable control room concepts.

If properly designed, control rooms, which are normally unused during portions of the day, will be configured as sophisticated editing facilities. This concept will be extended to other functions as well, such as dubbing, viewing and simple editing.

Studios with movable partitions, easily divided into larger or smaller production spaces, will assure better utilization of high-cost building space. With proper attention to acoustical isolation using recently developed techniques and materials, the concept of divisable studios may finally be realized.

The use of modular construction techniques will be maximized as a means of quickly upgrading a facility when new equipment is introduced. A concurrent task will be to find innovative ways to minimize the initial labor cost for installation. Automation and robotics will be utilized extensively as a means of obtaining productivity improvements. Program integration, signal routing, tape and materials handling and camera operations, as discussed before, will all be automated to the degree that it is cost effective and reasonable. Artificial intelligence will play an exciting role in the automation systems of the near future. It is now being widely recognized as the missing link between man and machine..and the applications for artificial intelligence will not be limited to our production systems, but will also play an important role in systems engineering, documentation and maintenance.

INDUSTRY PROGRESS BY 2002

The theme of the Conference is a forecast into the next century. Let me tell you how we see this at NBC.

We have two objectives in the design and implementation of NBC projects - maximizing productivity and, at the same time, improving quality. Quality improvements will come from the development projects whose aim is to save money.

A good example is the present M-II recorder program, which will eliminate all other video tape formats from our factory except for the very high end post production and graphics and the few recorders required for playback of source material on other formats, such as archival material. The implementation of M-II lowers costs, lends itself to automation and provides a quality improvement in ENG. It's a terrific example of what we're after. It gives us a win in spending, a win in quality, a win in flexibility and a win in operations.

In the next 15 years, we shall probably see the following innovations and changes in the television industry.

Changes to analog component systems in some parts of the industry and to digital component systems in most parts: this will evolve in most cases through the use of component "islands" of equipment, though some users may choose whole systems in components at one time.

NTSC will still be a major transmission medium, but will be substantially enhanced vertically, horizontally and temporally, both by improvements made in the broadcast chain and in the receiver.

Some form of HDTV will be introduced, which will probably be backward compatible to NTSC. This will certainly have a 5.33 to 3 aspect ratio, but how many lines it will have and whether one world standard or even one national standard will be achieved is temporarily an open question.

The means of getting this to the home may be a selection or combination of terrestrial broadcast, DBS and fiber. There will be a considerable move into high definition television program production - in particular, movies and specials - though what line standard this will be is not clear.

Cameras will be almost exclusively CCD; recorders will be almost entirely small format cassettes.

The signals reaching the home will generally provide much higher quality picture and digital stereo audio.

Fiber optics will not only find use in outside broadcast and hostile environments, but the restructuring of costs due to competition will allow us to use fiber within the broadcast plant.

There will be extensive introduction of automation in plants, including wide use of equipment with self-diagnostic capabilities.

Let me stop for a moment on that. We can see that manufacturers will have to provide fault diagnosis services by computer, telephone line and modem. The experience at NBC with the remote diagnostic capability built into the Ku Band satellite system clearly indicates to us that this capability must be extended into the rest of the factory as well.

Likely Changes In The NBC Operation

As discussed earlier vis a vis the M-II program, all playback functions will be fully automated in New York, Burbank and the Owned Stations. Recognizing, of course, the immediacy of news and sports, certain recorder functions will still be primarily manually operator-controlled.

Manual handling of tape in a fully automated system will be reduced to one time when the program material comes in-house, recorded or unrecorded, and a second time when it is removed for disposal.

Digital component systems from camera to the output of the plant will probably be the norm, with almost all CCD cameras, digital stereo audio from the audio console on, probably two formats of recorder covering a range from top quality studio recorders to small portable recorders for news gathering.

There will be substantial enhancements to the vertical, horizontal and temporal characteristics of the signal and some of the enhancements will require an improved receiver to benefit in the final displayed picture.

It is very likely that NBC will have some form of high definition television production and transmission capability. It will be 5.33:3 aspect ratio, but what line standard it will be I would not like to predict at this time.

CONTRIBUTION OF STANDARDS

Some may argue that the production of standards puts technology into a straightjacket, restricts manufacturers' inventiveness and design flexibility and thereby produces less satisfactory products. On the other hand, proponents of standardization will say that the interchangeability provided by standards encourages a much greater sale of products made to those standards. Standardization seems to have been a very positive factor in the type C one-inch video tape recorder and lack of standardization may have prevented the emergence of quarter-inch video tape recorders. In the consumer section of the television industry, lack of standardization may cause considerable problems in descrambling the various television signals received in scrambled form over the air. It could be argued by manufacturers that lack of standardization in half-inch video tape recorders has permitted greater inventiveness to be applied by Matsushita, Sony and others and has thereby permitted a better product to be produced than would otherwise be the case.

It should be clear to all of us that the path to the future requires a planned, coordinated and cooperative effort. Standardization is one of the most important parts of that cooperative effort. If standards are produced in accordance with good practices, in a timely manner, then all sections of the industry will benefit. For instance, as I stated earlier, now is the time to create a standard for the D-1 complementary ENG/EFP format.

The standards must, of course, define the minimum number of parameters necessary to allow interfacing between different manufacturers' products. The timeliness of preparation of a standard is also extremely important. If a standard is started too soon, there will be insufficient interest, not enough experts will participate and important issues may be missed in the standard. If the standard is produced too late, then de facto manufacturing situations will probably exist, compromises will have to be made and the optimum standard will probably also be missed. With the proper timing and the optimum rate of progress, manufacturers, users and the whole broadcast industry should benefit.

SUMMING UP

As you can see, there are many forces influencing the direction of our industry and we can predict many sequences of events depending on the interaction of those forces.

Today I have suggested how we, in general, and NBC, specifically, may use these forces to direct our future. However, no one can guarantee that there will be no changes - technical, political or economic - which will cause major revisions to our plans. When the astro-physicists reviewed the data from the Voyager-Jupiter flyby, it caused them to rethink their understanding of the universe. The Voyager-Saturn flyby produced data which caused further modification to their theory and the Uranus flyby caused yet another reshuffle.

What we can predict with a fair degree of certainty is that the future course of events will not be dull, but will be as exciting and challenging as any we might hope for.

I would like to leave you with one last thought. In my position as a Broadcast Manager with a large capital budget, I'm always being asked if technology is my friend or my enemy. I'm sure we feel that technology is our friend, but I believe that the real challenge over the next 15 years is how do we keep it from becoming our enemy.

Michael J. Sherlock is president, operational and technical services, at the National Broadcasting Co. (NBC). From November 1982 until his latest appointment, Mr. Sherlock was executive vice-president, operations and technical services, NBC-TV. He served as the chief financial officer for sports, with the title of vice-president, finance and administration, NBC Sports, from November 1979 until February 1982. For nine months before that, he had been executive vice-president, production administration, NBC Finance, with the responsibility for the unit managers, TeleSales, and program merchandise areas. From 1977-1979, he was vice-president of business affairs and administration, NBC News, where he was in charge of the division's business, planning administration, and finances. He came to NBC News from the Hertz Corp., where he served as vice-president, administration. He first joined NBC in 1960 in business affairs, rising to the position of director of business administration in 1973, before moving to Hertz. Mr. Sherlock holds a Bachelor's degree in business administration from LeMoyne College in Syracuse. He is an officer of the Center for Advanced Television Studies, a member of the Presidential Advisory Council of the Society of Motion Picture and Television Engineers (SMPTE), the International Radio and Television Society, and the Symposium Committee of the International Television Symposium and Teechnical Exhibition, Montreux.